Routledge Revivals

THE THEORY OF EVOLUTION
IN THE
LIGHT OF FACTS

THE THEORY OF EVOLUTION IN THE LIGHT OF FACTS

BY

KARL FRANK, S.J.

WITH A CHAPTER ON ANT GUESTS AND TERMITE GUESTS

BY

P. E. WASMANN

TRANSLATED FROM THE GERMAN

BY

CHARLES T. DRUERY, F.L.S., V.M.H.

Routledge
Taylor & Francis Group

First published in 1913 by Kegan Paul, Trench, Trübner & Co. Ltd.

This edition first published in 2018 by Routledge
2 Park Square, Milton Park, Abingdon, Oxon, OX14 4RN
and by Routledge
52 Vanderbilt Avenue, New York, NY 10017, USA

Routledge is an imprint of the Taylor & Francis Group, an informa business

Publisher's Note

The publisher has gone to great lengths to ensure the quality of this reprint but points out that some imperfections in the original copies may be apparent.

Disclaimer

The publisher has made every effort to trace copyright holders and welcomes correspondence from those they have been unable to contact.

A Library of Congress record exists under ISBN: 28001429

ISBN 13: 978-0-367-15035-8 (hbk)
ISBN 13: 978-0-429-05465-5 (ebk)

THE THEORY OF EVOLUTION

IN THE

LIGHT OF FACTS

Nihil Obstat.

Sti. Ludovici, die 19 *Sept.*, 1912

Josephus Wentker

Censor Librorum.

Imprimatur

Sti. Ludovici, die 22 *Sept.*, 1912

✠ Joannes J. Glennon

Archiepiscopus Sti. Ludovici.

THE

THEORY OF EVOLUTION

IN THE

LIGHT OF FACTS

BY

KARL FRANK, S.J.

WITH A CHAPTER ON ANT GUESTS AND TERMITE GUESTS

BY

P. E. WASMANN

TRANSLATED FROM THE GERMAN

BY

CHARLES T. DRUERY, F.L.S., V.M.H.

VICTORIA GOLD MEDAL OF HONOUR IN HORTICULTURE

WITH 48 ILLUSTRATIONS

LONDON

KEGAN PAUL, TRENCH, TRÜBNER & CO. LTD.

B. HERDER: 17 SOUTH BROADWAY, ST. LOUIS, MO.

1913

PREFACE.

The object of the present work is to throw some light on the theory of Descent. Among many of the students of nature of the present day we perceive that greater and greater contradictions arise between the actual results of their technical work and that which they put forward as 'postulates' of the theory of Evolution. Our object is to deal with this. The 'certain' or the 'probable' should be separated from the pure 'postulates' and the actual area of elucidation of the hypotheses of Evolution be thereby clearly defined. The chief postulate, the origin and development of the animals from the plants, is dealt with fully.

If the area of elucidation be exactly defined, then, and then only, can it be a question of a truly scientific attempt at explanation. The best known systems, those of Darwin and of Lamarck, are tested by their premisses and method; but only their *specific* doctrines are taken into consideration.

In the formation of a reliable hypothesis it was less important to present a complete collection of all observations than to show clearly, by separate examples, the demonstrative force and extent of the argument concerned.

A few chapters have been already published by the Author, but in another connection—in the 'Lehrbuch der Philosophie,' 3rd ed., of Father Lehmen, S.I. (published by Father Peter Beck, S.I.). Even these, however, have been partly rewritten and extended.

My brother in the order, Father E. Wasmann, has been good enough to place at my disposal a contribution from his special sphere of study. Further contributions regarding similar questions he will himself publish in the fourth edition of his work, 'Die Moderne Biologie und die Entwicklungstheorie,' which will shortly appear.

THE AUTHOR.

BALKENBURG.

CONTENTS.

CHAPTER II.

SUGGESTIONS FOR RELIABLE HYPOTHESES OF EVOLUTION.

ILLUSTRATIONS.

THE THEORY OF EVOLUTION IN THE LIGHT OF FACTS.

SECTION I.

GENERAL (PALÆONTOLOGICAL) BASES OF THE THEORY OF EVOLUTION.

CHAPTER I.

ORIGIN OF THE PROBLEM AND DEFINITION OF QUESTIONS INVOLVED.

§ 1. *Origin of the problem.*

A STUDY of the organic world in which we live demonstrates clearly that the animals and plants do not show a confused admixture of forms but display a separation into groups which can usually be defined with sufficient certainty. We find individuals which among themselves are fairly alike in all characters and which under normal circumstances can also only be perfectly fertile by intercrossing. The whole of these individuals which are so related to each other form a so-called *systematic* species. The Wolves, for instance, form such a species (*Canis lupus* L.). That by the word ‘wolf’ we represent a defined type of animal which cannot be confounded with another is shown by the pictures in our school-books and primers by which the children are made acquainted with this dreaded animal. A single picture is there indicated with the name of 'The Wolf' beneath it, and yet it is expected that the child will recognize by this single picture all wolves which it might ever see in any zoological garden.

Similar animals, but not wolves, are the Dogs, wild and tame, and the Foxes. The make of the paws, the toes, the entire habit and mode of life (Carnivora) are very similar; on the other hand they differ entirely in all these characters from the Sheep, Goats, etc. On the strength, therefore, of the said similarities the wolf, dog, fox, etc., are placed in a higher systematic unit, viz. in the dog family of the Canidæ. Particularly in the shape of the jaw (carnivorous jaw) all dog-like animals resemble the Bears, the Marten, and others; all these animals can therefore be united to a further group, viz. that of the land Carnivora (*Carnivora fissipedia*). According to the similarities which even then remain between many groups, the allocation of animals to higher units progresses, which naturally then become ever more and more general and comprehensive. The land Carnivora are united with the aquatic Carnivora (*Pinnipedia*) to form the order of Carnivora; then with all sucking animals to form the class of Mammalia; and finally with all animals which possess a skeleton formed of jointed vertebræ, to form the tribe of Vertebrates.

It is now a question whether this graduated division, which renders it possible for us to unite all animals and plants into a few tribes and classes, is only the expression of a general plan which the Creator wished once and for all to realize, or whether this similarity rests really upon actual relationship (blood-relationship). We can easily come to the latter conclusion, since in point of fact we observe that the greater the resemblance between definite individuals so much the closer is the true blood-relationship. Children, as is often said, resemble the parent like 'their very image,' and brothers and sisters are, as between themselves, of the greatest similarity. Negroes, Indians, Chinese are also 'true' men and resemble the Europeans, but the reason of their perfect racial resemblance, which is shown by the Negro, we seek and find again in their descent from black parents, in their close or more distant blood-relationship. Ought we or must we therefore also, in order to explain that Europeans, Negroes, Indians, etc., are all men, accept a common descent for all and therefore an actual blood-relationship, even though it be a cousinship 1000 or 10,000 degrees removed?

The same question can also be raised in view of the similarity of the systematic hierarchy established for animals and plants. Is every resemblance, be it ever so trifling—as, for instance, the possession of a vertebral column—the expression of an actual blood-relationship, or—what is the same thing—the descent from common progenitors in the dim and distant past?

The question cannot obviously be determined *a priori* nor even by the direct observation of the animal and plant forms of the present day. We observe, indeed, that blood-relationship (descent from the same parents) never establishes a more general or more extended similarity than the 'specific similarity': that is, the most perfect similarity which we in fact know of. Progeny of the same parents never depart so far from each other or even from their nearest 'relatives' that we rank them of different species, and therefore we must create a 'race' for all of them in order to unite them generally.[1]

Upon this observation the opinion is based that the specific similarity is above the expression of actual relationship—which specific similarity is always, so far as we can observe it, transmitted unchanged as regards the essential distinct characters in the process of reproduction, i.e. remains constant: that would be the doctrine of *specific constancy.*

The greater or less similarity also with other animal types was well recognized, but since it was seen how they continued to exist together but separated, and crossed either never or unwillingly or unfruitfully, it struck no one that similarity, carried further, could, and must, be based on descent.

Already the attention of some investigators had been aroused by certain petrifactions. Many of these showed clearly the form of mussels, fish, leaves, etc., but they often appeared quite different from the corresponding animals, etc., with which man was familiar. Sea mussels and sea fish were also found far inland or

[1] We choose this mode of expression purposely, which moreover cannot be contested, in order to avoid long explanations as to the meaning of 'species.' Compare Wasmann: *Die Moderne Biologie und die Entwicklungstheorie,* chap. x.

even on the mountains. Endeavours were made to explain this in very varied ways, either by acceptance of oceanic floods which carried the sea animals inland, or it was disputed that they were really the remains of pre-existent life.

To this end a peculiar force in stone was conceived—*nisus formativus*, or petrifactive force—which, with or without the assistance of the stars, imitated organic forms from inorganic materials.

This opinion, in the sixteenth and seventeenth centuries, was the ruling one despite the better knowledge of some eminent men.

Leonardo da Vinci, for instance, would have nothing to do with such enigmatical working of the constellations.

Since, however, it was considered that the earth and its organisms had been created as they then were, the *nisus formativus* appeared to be the more acceptable explanation. Consequently the 'petrifactions' were not at all regarded as the remains of organisms which at one time had really lived; and of a connection with the living animals and plants of to-day no one then thought. The opinions of the time are shown, for instance, in A. Kircher's 'Mundus subterraneus' (Figs. 1–4).[1]

Since, however, the number of such 'figure stones' discovered constantly increased, grave doubts began to arise against the *nisus formativus* theory, especially

[1] II, Amstelodami 1665, c. 9. Kircher also stated that many of these *lusus naturæ* might have originated through hollow spaces in the earth becoming filled with mud.

since it apparently had ceased to act. The conclusion became enforced that the petrifactions should be considered as remains of actual organisms which, however, certainly had, in the opinion of the time, nothing to do with the still existent forms of animal and plant life; they were extinct, i.e. types of life which had been annihilated by the one universal flood, the Noachian Deluge.

FIG. 1.—LUSUS NATURÆ.

Obviously fancifully completed. Fig. 2 is based on a mussel (Inoceramus). (*After A. Kircher.*)

The most formidable upholder of this view was undoubtedly J. J. Scheuchzer (1672–1733). He took up arms courageously against the current ideas of aerial spirits (*Archaei*) which bury themselves in the soil and stones and so shape organic forms. 'Such idols,' he says in his book 'Homo diluvii testis,' 'must be overthrown and destroyed, not so much by subtle philosophy and all sorts of brain whims, but by presentation and observation of Nature's bodies themselves, and such resulting consequences as even the simplest may seize on and understand. Nature must be her own advocate, and wisdom, though unstudied, must be the judge.' He also expressed himself strongly against the *lusus naturæ*. In his paper *Piscium querelæ et vindiciæ* (1708) he makes the Fishes raise objections that they are not considered as the original parents of the present fishes but are regarded as of 'mineral stone and marl births.'[1]

In his splendid work 'Physica sacra'[2] he goes through the separate groups of animals which were destroyed by the Flood. His copperplates are excellent, his added verses less so.[3]

With regard to the origin of the mountain ranges he had also some wonderful ideas. When the *Gemüsz* (vegetable debris), into which the earth's strata had been changed by the Flood, dried again, the crust burst and there were heavings and sinkings. That earthquakes and the like had been able to form the Alps was, in his idea, a 'lame opinion.'

[1] K. v. Zittel: *Geschichte der Geologie und Paläontologie bis Ende des 19 Jahrhunderts*, Munich-Leipzig, 1899, 24.

[2] I, p. 61, Augsburg und Ulm, 1731.

[3] Since he found remains of all animal forms, he deduced that all animal life was annihilated, and, from that, that the Deluge was universal:

'Since all that lived and moved therein was drowned
'Tis clear the Flood prevailed the whole world round.'
(p. 64, Translation.)

'The man of evil luck's remains likewise from out the ground
Have now been dug, and for it many reasons have been found.'
(p. 66, Translation.)

This 'man of evil luck' ('a disturbed skeleton of an old sinner') eventually proved to be the skeleton of a gigantic Salamander (now in Haarlem).

The Noachian Deluge was made responsible for

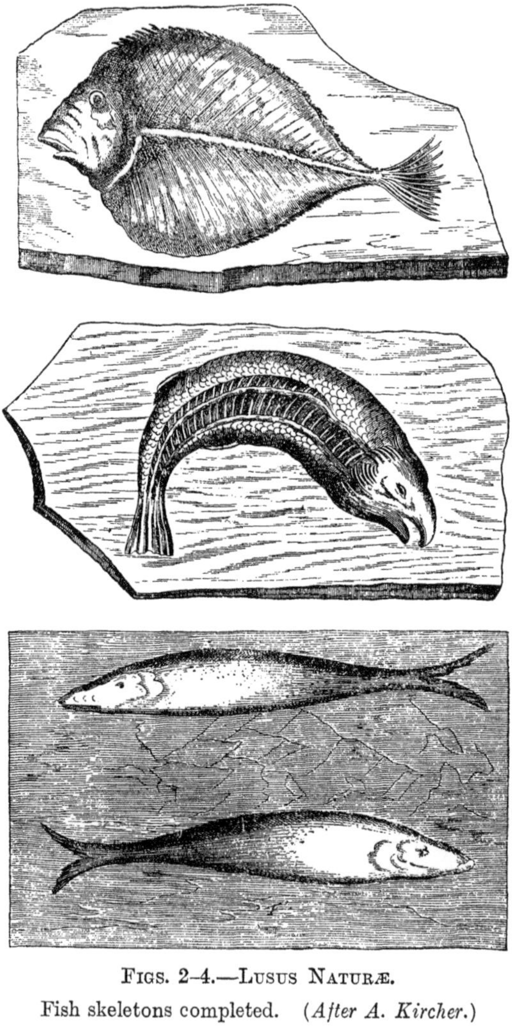

Figs. 2–4.—Lusus Naturæ.

Fish skeletons completed. (*After A. Kircher.*)

everything. That the earth's strata could be formed by quiet deposit of flooded materials such as sand

and mud, as now happens, he had no conception. As regards the causes of elevation and depression and the significance of volcanic catastrophes the proper understanding failed him entirely.

But this point of view likewise failed to be permanently maintained. It was gradually recognized that the separate and sharply defined strata concealed varied fauna and flora in their depths. There must, therefore, have happened several such mighty floods, or catastrophes of other kinds—the living world was annihilated repeatedly and arose again as often. Since, however—precisely because they often varied greatly from each other—no one conceived the thought that the separate successive organic forms might derive their origin from each other by descent, nothing remained but to explain each new organic world by a new creation or successive creation.

It was not every investigator who understood by 'new creation' a creative act of God. Cuvier, who often used the word, assumed that after the destruction of a defined fauna a new one immigrated from somewhere, so that in this way a new creation, i.e. new creatures, took the place of the extinct ones. Cuvier, however, had no idea of a genetic connection (by descent) between the successive and varied organisms.[1]

Cuvier's pupils, d'Orbigny, d'Archiac and others, carried their master's teaching further: d'Orbigny[2]

[1] Compare Chr. Depéret: *Die Umbildung der Tierwelt*, Stuttgart, 1909, 10 (German translation by R. N. Wegner of the French work, *Les Transformations du Monde Animal*, by Chr. Depéret).

[2] *Cours Élémentaire de Palæontologie Stratigraphique*, II, Paris, 1849, 251.

established twenty-nine quite different creations, each quite independent of the other, which owed their existence rather to twenty-nine different acts of creation. This happened so late as the year 1849.

'A first creation,' said d'Orbigny, 'shows itself in the Silurian formation. After the entire destruction of this by some geological cause, and after expiry of a considerable period, there occurred a second creation in the Devonian formation and thereafter twenty-seven successive and different creations have repeopled the whole earth with plants and animals in connection with the geological cataclysms which had previously destroyed all living nature. These are facts, certain but incomprehensible facts, which we confine ourselves to stating without attempting to pierce the super-terrestrial secret which enfolds them.' [1]

We point out that such eminent men as Cuvier (the father of Palæontology) and his pupils found the separate successive organic worlds so fundamentally different from each other that they conceived absolutely no idea that the later ones could arise from the former. Therefore no one either can or should doubt any longer that the organisms really appeared different in the past than they do now, and, moreover, the comparative difference is the greater the older they are. Both facts are established by investigators who had no interest whatever that it was so; rather did it confuse them to be thus confronted with 'incomprehensible facts.'

[1] Depéret-Wegner: *Die Umbildung der Tierwelt*, p. 15.

D'Orbigny might well feel that the acceptance of twenty-nine complete destructions and complete new creations involved something unworthy of credit. The number was also arbitrarily chosen. If the organisms were always so unchangeable as they appear to be now, then, as a consequence, new creations must be accepted and also the time between each two catastrophes; since many animal types show, on one and the same geological strata-system, such changes as the theory of 'specific constancy' cannot admit.

Why, furthermore, do the new creations so often approach the extinct forms so closely that in many cases the conclusion forces itself upon us, that the immediately successive creation must be a simple further development of that immediately preceding? That is particularly the case when in the immediately preceding creation the organisms already show clear traces of changes in a definite direction, which need be only increased, i.e. further differentiated, in order to perfectly explain the appearance of the immediately succeeding animal and plant creation. Furthermore, the geologist in the meantime had recognized, by more exact observation of the geological formations and denudations of the present day, that the fossil-bearing strata could not have originated catastrophically. At the most such an explanation can only be applied to particular local changes. Lyell especially has the merit of this more natural comprehension of the terrestrial formations ('Principles of Geology,' 1st ed. 1830–38).

It will, therefore, be willingly conceded that the

'catastrophic' and 'creation' theories do not afford the most natural explanation of the history of life upon the earth. Another hypothesis was, however, still impossible so long as the absolute unchangeability of the organisms was adhered to.

Already, in Cuvier's lifetime for the first time, serious attacks began to be made on the 'constancy' theory in order to permit the possibility of another signification being given to palæontological facts, and particularly to establish an actual connection between the present and fossil organisms. Lamarck and Geoffroy St. Hilaire had already, decades before the appearance of d'Orbigny's 'Cours élémentaire,' put forward theories on quite new and unheard-of lines. But their ideas found no proper response among their contemporaries. It was reserved for Charles Darwin to obtain for the new doctrine a general acceptance; in what sense that should be properly understood we shall speedily see.

All agreed that the living and fossil animal and plant forms are connected genetically, i.e. by descent. Thus the many new creations became superfluous. These were, indeed, only insisted upon because that connection was denied. The difference established by palæontology of offspring between themselves and from their ancestors finds its explanation in the variability of the organisms, which, in the sharpest contradiction to the 'constancy theory,' became regarded as a fundamental quality of all living things. Since palæontology furthermore presented some evidence that the difference of form of the progeny was accompanied by a certain degree of improvement in the sense of development (evolution) of previously existing points, the new hypotheses were termed 'evolutionary hypotheses.' Evolution, therefore, implies more than change of form. The French scientists, especially palæontologists, frequently use up to the present the word 'transformism' or 'transformation': thus the excellent work of Ch. Depéret (1909) bears the title of 'Les Transformations du Monde Animal,' the German translation by Wegner that of 'Die Umbildung der Tierwelt' (a literal translation).

§ 2. *Questions involved.*

If we would now come to a decision regarding the history of organic life, then, according to the above, the following questions must be put and answered in the given sequence :

(1) Is there a general genetic connection between the organisms of to-day and the fossilized ones ?

If yes, then we accept without further argument a change, a transformation of the organism, since the old animal and plant worlds appeared different from the younger and present ones.

(2) Is this change, this transformation, connected with a higher development ?

This question cannot possibly be decided unless we know what palæontology teaches us :

(*a*) With regard to the succession of definite well-characterized groups (types).

(*b*) With regard to the genetic connection of the later (higher) groups with the earlier (lower) ones.

(*c*) With regard to the eventual changes *within* such groups.

If the so-called higher groups follow the lower but do not arise *through* them then we have no higher evolution in the narrow sense of the word, because then the entire transformation effects itself within the same plane of organization, for example, within the same order.

If we come to a definite conclusion regarding these three questions, then, and not before, can we think of formulating an hypothesis which will explain or make comprehensible that which has been demonstrated or accepted as probable.

Whoever accepts the genetic connection of all living things[1] must seek for an 'hypothesis of descent,' and since with the genetic connection a transformation is essential, also an 'hypothesis of transformation' must be sought for. Whoever in addition accepts a higher development needs an 'hypothesis of evolution' if he will explain everything. In point of fact the expressions 'evolution,' 'transformation,' 'descent' become less strictly separated. Furthermore the terms

1 It is naturally here an entirely general question whether we can accept, as a rule, a connection between present types and those of earlier times of different appearance. To inquire which organisms thereby come under consideration is a matter for the palæontologist. For man, by reason of his essentially higher elevation over the whole of the rest of living beings (the spirituality of the soul), there must be accepted, in any case, a special intervention of God. For man, if he be taken as a whole, no beast can be regarded as ancestor, however highly it may be developed.

used to indicate the scientific value of the endeavours at explanation, 'hypotheses,' 'theories,' or even 'doctrines,' are constantly changing. We believe that the expression 'theory of evolution' best fits the present position of the scientific world. We will, however, use the other expressions as opportunity demands.

All hypotheses must, according to the scientific principles generally followed, start from the present observation of the organisms concerned, and may not, as a basis of explanation, combine therewith anything which is contradictory to the certain teachings of other sciences, especially that of philosophy. They must furthermore state precisely what they really claim to explain and therefore must above all things adhere to the results of palæontological research.

§ 3. *Conclusions from Chapter I.*

As is seen from the above, it is not a question in this chapter of a positive proof of a descent of the present organic forms from earlier ones. A genetic connection is only accepted in order to avoid what is regarded as an unnatural presentation of repeated annihilations and new creations of entire organic worlds. The difficulty in which investigators were placed by the catastrophic theory has become greater and greater since fossils were discovered in the deepest strata, while even within the compass of single formations changes were observed in certain animal and plant groups, and particularly since geological processes of the present time have been ascertained which can fully explain the origin of fossiliferous formations without any great catastrophic cause at all.

Neither philosophy nor theology compels us to accept any such hypothesis of creation as involves, for instance, that God destroyed all fishes twenty-nine times (d'Orbigny) and created them anew twenty-nine times, the newly created differing generally more or less from the destroyed ones while some were replaced almost unaltered.

The only objection which could be raised was the following: It is not necessary to assume that the present forms descend from others of different appearance, but from like or very similar forms which previously existed near the differing ones, the remains of which have so far not been discovered. In that case the 'catastrophic' theory would also be superfluous, and despite it a so thorough transformation of the organisms should not be accepted as established, since experience now appears to contradict it.

Next there should be remembered the words of d'Orbigny regarding the 'complete' divergence of the earlier forms from those of to-day, which were not shaken even by some examples of striking constancy within very narrow limits of relationship, which have continued from the oldest periods to the present. How the unchangeability of living organisms is to be regarded we shall consider later on. For the rest it is inexplicable how precisely those differently formed animals and plants in the strata accessible to us have been preserved, while those similar to the present ones, but which mostly lived together with them under the *same conditions*, are generally rare.

The objection that most fossils lie at the bottom of the sea and cannot be examined by us at all, does not help us out of this difficulty. Thereby we could 'demonstrate' everything.

We are therefore inclined to accept a connection by descent between the present and the fossil organisms, because this assumption is the more natural provided that the observation of the present organisms does not *exclude* such an hypothesis.

The object of earnest scientific investigation would be to examine more closely the present forms of organic life as regards their varietal capacity and to follow up the evolutionary series of the separate groups and no longer to maintain in a merely general way a genetic connection as a postulate. This work, especially the palæontological side of it, has been, however, only resumed in recent years on really unprejudiced lines. To this we owe the really natural further development of the question of descent which for some decades had been interrupted by hypotheses of a merely general character.

CHAPTER II.

RESULTS OF PALÆONTOLOGICAL RESEARCH INTO THE EVOLUTION OF ORGANIC LIFE.

A.—PALÆONTOLOGICAL RESULTS.

§ 1. *Brief purview of the chronological succession of the larger animal groups.*

FOSSILS are exclusively found in the so-called sedimentary deposits which form the upper part of the earth's crust. Sediments (sedimentary or secondary formations) are such rocks, nearly always appearing as layers, which by deposit of gravel, sand, or mud (or by decomposition of dissolved minerals—salts for instance), are formed in the greater water-basins (in the sea or fresh-water lakes). The foundation of investigation of the origin of organisms is therefore an exact determination of the age of the sediments. It is only when it is known which stratum or layer (a) is older or younger than another (b) that we can also know which organisms are older or younger than others accordingly. This determination of the age of the earth's strata is, however, a very difficult matter, and the course of evidence which led to the generally recognized arrangement of the four (or five) groups of formations, is not far removed from a vicious circle, especially when we consider the mode of expression used by many authors.[1]

The uncertainty which exists, if we accept the usual division into separate groups determined by age, may be judged by the following short consideration: If it be no longer a question whether the organisms generally vary, but rather how they transformed themselves, then it is not sufficient to compare formations differing considerably in age, but those immediately following each other must be known, since it is only when it is known which formation was the next to be deposited, that the further fate of a definite organic group can be properly followed up without a break. Then next younger, which we will call 'b,' need not necessarily be deposited over stratum 'a' which has just been formed, but may originate in quite another region. The stratum 'a' can, for instance, become dry land by the retreat of the sea in which it was formed. The sea itself departs, together with its organisms, which hitherto had been buried in 'a,' to some other region and *there* deposits the successors of the organisms buried in 'a.' If there be no means of recognizing this next younger deposit, or if it be again covered by the sea, then nothing can be said regarding the evolutionary progress of such a group, or at least there exists a gap. Then it may happen that the animal groups, which we learnt to recognize in the strata complex 'a,' in that deposit ('c')

[1] Compare the methods of age-determination by E. Kayser (*Lehrbuch der Geolog. Formationskunde*, Stuttgart, 1909, p. 2); and also the Introduction of M. Neumayr, *Erdgeschichte II*.

in which we meet them again for the first time, show an entirely peculiar appearance, so that at the first glance no one would think of any connection with the fauna of 'a.' If, however, the groups of formation 'a' show a distinct tendency to vary in a definite direction, and if from a comparison of the fauna of 'c' with that of 'a' it is seen that the heterogeneousness consists in a great but apparently interposed increase of just those variational tendencies evinced in 'a,' then it may be assumed with great probability that the organisms in 'c' are the modified offspring of those of 'a.' The intermediate links lie buried in 'b,' and this formation is possibly now and has been for a long period covered with water and therefore inaccessible to us. The same process can also be repeated for 'c.' One of the fauna of 'c' approximating thereto, but greatly modified, may for instance only be found again in an obviously much younger formation, say in 'f,' and this may be in North America while 'c' may be in Europe. (The strata 'd' and 'e' lie perhaps under water or have not been investigated.) If we accept a connection between 'f,' 'c,' and 'a,' then we have obviously only important outposts as it were in the march of evolution of a particular group, and perhaps also a general indication how the outpost 'f' can have been derived from 'c' and 'a,' but without any precise knowledge of the process involved.

Only in a few cases: as, for example, the same sea in which the formation 'a' was deposited, may, in a short time, return to its old position (sea oscillations), and consequently the same organisms also return, so that in the stratum 'b' which is formed after the return we have before us the immediate descendants of 'a.' Several such cases can be recognized with sufficient certainty. Then by comparing 'b' and 'a' we arrive under certain circumstances at a clear insight into the mode of variation and its rapidity, etc.

If the fauna of 'a,' or a group of the same, should not, generally speaking, reappear, and is no longer seen at the present day, then it is 'extinct.' How and when it became so, we are so far ignorant.

It is therefore seen how difficult it is to make clear the process of evolution for a definite group. Many geologists entirely despaired of the possibility of so exact a definition of the ages of the formations as was needful to that end. Incomplete, very incomplete indeed, must our knowledge ever be.

As has been stated, we can grant that for the great geological formation system the relative longevity has been ascertained with sufficient certainty. Beginning with the youngest, we have arrived at the following generally used table of the geological periods:[1]

Formation Groups.	*Formations.*
Quartiary . .	Alluvium Diluvium (Pleistocene)
Tertiary	Pliocene Miocene Oligocene Eocene

[1] For more exact indications and subdivisions *vide* Kayser: *Lehrbuch der Geolog. Formationskunde*, p. 7.

Formation Groups.		*Formations.*	
Secondary . .	Chalk	Upper Chalk	with many grades
		Lower Chalk	
	Jura	Malm	
		Dogger	,, ,, ,,
		Lias	
	Trias	Keuper	
		Mussel Chalk	,, ,, ,,
		New Red Sandstone	
Primary	Dyas (Perm.)	Permian Limestone	
		Old Red Sandstone	
	Carboniferous		
	Devonian		
	Silurian		
	Cambrian		

To these are to be added now the pre-Cambrian (Algonkium). Under these lie gneiss and mica slate.

The first fossiliferous formations are the so-called pre-Cambrian. The primary rocks (gneiss and crystalline slate), upon which the pre-Cambrian sediments lie, conceal no organic petrifactions of any kind at all.[1] It must therefore be accepted that, in the seas in which the oldest sediments were deposited, life really appeared in the first place. That must have happened very long ago, since, if we imagine all the formations superposed on each other, the total would be of a thickness of about 200 kilometres—about 120 miles.

[1] With regard to the alleged fossils of the Primary rocks see Kayser: *Lehrbuch der Geolog. Formationskunde*, p. 21. The freedom from fossils of the Primary rocks 'is only temporarily shaken'—at first by the Eozoon Canadense, which was recognized as serpentine excrescences. Since then other assumed traces of organisms have deceptively appeared, but the Primary rocks must be again, as previously, regarded as entirely free from fossils; 'also the presence of lime or graphite does not imply organic life, since both have been proved to be able to originate also on inorganic lines.'

(1) *First appearance of life in the lowest (oldest) sedimentary formations.*

The oldest well-preserved fauna (of plants nothing has survived but a few marine algæ) is that of the Cambrian system. It occurs in striking abundance and extent. 'The majority of the important groups of invertebrate animals were already clearly differentiated,' and the Foraminifera, Sponges, Corals (and Medusæ), Brachypods, Snails, Cephalopods, and Arthropods were present. These groups were partially again split up into many species and genera: the Trilobites (Crabs) formed fifty genera with 150 species; the Echinidæ occur in three types; the Cephalopods, the 'highest type of the Mollusc class,' are already represented (Orthoceratidæ). The Crabs proper (independently of the Trilobites) form two well-separated groups (Ostracoda and Malacostraca).[1]

On the other hand no remains of Vertebrates have yet been found.

In the pre-Cambrian formations, which have nearly everywhere experienced great metamorphic changes, fossils are found only now and then. According to the latest investigations undertaken by J. Walther[2] on the spot (in California, Scotland, and Norway, the sites of the most important discoveries in pre-Cambrian strata) there have been found, as the most ancient traces

[1] Depéret-Wegner: *Die Umbildung der Tierwelt*, p. 233. Kayser: *Lehrbuch der Geolog. Formationskunde*, p. 75.

[2] *Ueber Algonkische* (=*pre-Camb.*) *Sedimente,* Naturw. Rundshau, 1910, p. 158.

of life, worm-tubes, trilobites, brachypods, and snails (among them a genus which still exists—Pleurotomaria of two species), these representing already fairly developed organisms. Other finds in Brittany are dubious. But obvious limbs of starfish (Crinoidæ) occur.

The pre-Cambrian fauna appears therefore to be about as perfect as that of the Cambrian formation itself.

Conclusions from (1):

(*a*) The first organisms appear together, not successively, in types or groups clearly separated from each other.

Most of the invertebrate classes are found.

Many forms, it is true, carry the impress of simplicity (' clumsiness ' it might be termed), like the Crabs; the Snails are still small and but slightly typical; the Cephalopods are only represented by puny forms with a flattish shell[1] (in contrast to the manifold crooked and ornamented shells of the later representatives of this group).

(*b*) Only one group (Trilobites) shows already a very profuse branching into divergent differentiated forms. Fifty genera and about 150 species.

(*c*) Monocellular organisms are preserved—Foraminifera—but they by no means form the chief component of the primary fauna (as the terrestrial evolutionary hypotheses demand); ' from the beginning of animal life we are already infinitely far removed '[2] i.e. the ' beginning ' of life had not been thus imagined.

[1] Kayser: *Lehrbuch der Geolog. Formationskunde*, p. 80.

[2] E. Koken: *Die Vorwelt und ihre Entwicklungsgeschichte*, Leipzig, 1893, p. 82. This remark of Koken's is to be understood from the standpoint of certain hypotheses of evolution which prescribe a fauna of entirely different character.

(*d*) As regards the origin of the groups already existent in the Cambrian and pre-Cambrian period we shall never know anything certain, not even if other organisms really preceded them, as is demanded as a postulate by the extreme evolutionary hypotheses.[1]

In the meantime is the pre-Cambrian or Cambrian fauna to be scientifically regarded as the real original fauna ? The groups already existing show, in many representatives, about the medium height of organization of the animals of the present day. Of a fauna in its entirety, of a lower grade and still older, there have been in any case no remains discovered (see *c*). If such really existed we should hardly ever be likely to learn something from it (see *d*), since the formations which are older than the pre-Cambrian—nay, even these themselves for the greater part—are throughout so greatly metamorphosed that all and any fossils which they might have contained would be destroyed.

(2) *The form of animal life in the post-Cambrian formations.*

The justification for uniting several sedimentary formations in a single group (formation, e.g. Silurian, Devonian, etc.) is found in certain peculiarities which we see recur in the history of organic life in what

[1] Depéret-Wegner : *Die Umbildung der Tierwelt*, p. 312. 'From these facts should it be concluded that we must for ever desist from hoping to solve a problem so passionately discussed as that of the commencement of life on the earth, or at least to be able to follow it further back ? Unhappily it must be granted that that is the most probable prospect before us.' (This remark shows also that a so highly developed fauna was not anticipated.)

might almost be termed rhythmic order. These peculiarities are the first appearance of new and higher types which were previously absent (larger groups, e.g. classes or families), vigorous development of some of the already-existing types, and the decay and disappearance of other and previously very varied ones.

The Cambrian or pre-Cambrian formations show in a general way organisms for the first time; they are sharply defined according to depth. None of the classes of Invertebrates which existed were as yet freely divided into genera and species, they show no specialized adaptation to the various environments (i.e. but few families, genera, and species); the Trilobites alone form an exception.

The individuals are still small and simply constructed; the Cephalopods, for instance, have simple straight shells in contrast to the highly complicated, curled, and ornamented ones of the Cephalopods of later formations.[1]

All at once there appear, in a definite series of strata, the first fishes, i.e. the first representatives of the family of Vertebrates; the first land plants also appear.[2] The Corals, Starfish, and Graptolites (related to the Bryozoa, but long extinct), which were very rare in the Cambrian formation, become more numerous.

The Trilobites develop more abundantly, the more

[1] Good tabulated illustrations (one short and one in detail) of the chief groups of animals and plants from these three points of view are given in H. Credner's *Elemente der Geologie* (1902), pp. 363, 365.

[2] We remark once and for all that 'first appearance' is intended only to mean 'fresh found.' Perhaps the two are synonymous; perhaps not.

perfect forms with well-developed eyes and with the faculty of rolling themselves up, obtain preponderance; the Cephalopods, certainly only a branch of the Nautiloids, become plentiful and of many types; the Ammonites are almost entirely absent—their time has to come.

The shells of the Orthoceras forms are somewhat convoluted.

The Crinoids (rare in the Cambrian formation) become very numerous; to them are added, but at first sparsely, the two new classes of Echinidæ, the sea stars, and sea urchin.

The Bryozoa are also noticed, but they are far from having the importance they later acquired.

In short, we understand why—in view of the first Vertebrates and land plants, in view of so many new orders and families appearing within classes already existent, and in view of the great number of families, genera, and species into which other classes and orders, as it appeared, simultaneously and surprisingly quickly ('explosively') split themselves up—a new 'creation,' the Silurian, can be spoken of.

We cannot go through the formations separately, but confine ourselves here once more to the determination of the chief results which are necessary to further explanations.[1]

[1] E. Kayser (*Lehrbuch der Geolog. Formationskunde*), after discussing the separate formations, gives a good palæontological purview, to which the reader is referred. K. von Zittel's *Handbuch der Paläontologie*, Munich and Leipzig, 1876–1893 (5 vols.), goes more into detail. Regarding the history of the Vertebrates the best information is given by E. Frieherr Stromer v.

On the whole the higher classes of vertebrates in the post-Cambrian formation follow the lower, and often, within the classes, the higher forms follow the lower. The latter applies also in many cases to the Invertebrates.

The vertebrate classes as now represented are, beginning at the lower, the Fishes, Amphibia, Reptilia, Birds, and Mammalia.

The Fishes appear for the first time in the Silurian formation, and are divided into three different groups : these are the Sharks, which rank as low grade ; the other groups were armoured, clumsy forms which subsequently disappeared.[1]

The Amphibia appear in the upper Devonian.[2]

The Reptiles, certainly strange-looking forms, show themselves for the first time in the upper Carboniferous formation ('Sauravus').[3] In the Permian there appear two orders of Reptilia which have died out except one (genus *Hatteria*) which now exists in New Zealand and has always been regarded as a stranger in our animal world. Quite recently came the discovery of three great groups of 'well-developed land reptiles' in the Russian Permian formation.[4]

Reichenbach's *Lehrbuch der Paläontologie* (Naturwissenschaft und Technik in Lehre und Forschung), Leipzig-Berlin, 1909, Part 1. J. Bumüller handles the question on briefer lines (*Die Entwicklungstheorie und der Mensch*, Munich, 1907, p. 50). This excellent and inexpensive work is highly to be commended.

[1] E. Kayser, p. 138.

[2] Depéret-Wegner : *Die Umbildung der Tierwelt*, p. 229.

[3] *Ibid.*

[4] *Ibid.* p. 230.

The forms which most nearly approach the chief reptilian order of the present day—turtles, crocodiles, and serpents—appeared later.

The Birds are known to us through the two Archæopteryx from the Solnhof Slate (upper Jura). Remains of birds, as generally of all land animals, can naturally be but seldom preserved. The two Archæopteryx tell us practically nothing at all of the history of the Birds.

The Mammalia appear, as a class regarded generally, for the first time in the upper Trias [1] and in forms which nearly approach the lowest orders of the class of Cloaca and Marsupials.[2]

The higher orders appear later, but then certainly and simultaneously and partly in the most differentiated forms such as the whale (Cetaceæ) bats, and Proboscidæ (Tertiary in the Eocene period).[3]

Conclusions from (2): The higher classes of the Vertebrates appear after the lower (Birds ?); within the classes also the higher orders appear later than the lower.

Many groups (it might perhaps be said of all, were the evidence more perfect) show always at their first appearance already a division at least into some higher

[1] The Mammalia were therefore older than the Birds, the first remains of which come from the later Jura, if the Archæopteryx may not be regarded as shattered examples which have reached us altogether by chance. Possibly the birds are much older.

[2] Marsupials and Cloaca are primitive forms because, since with them the development of the embryos is—entirely with the Cloaca, or mainly, as with the Marsupials—extra-uterine. With the higher orders a placenta is formed, and the development is entirely inter-uterine.

[3] Further particulars concerning the classes of Vertebrates are to be found in G. Steinmann's *Die Geologischen Grundlagen der Abstammungslehre*, p. 203.

systematic categories. It is often observed that such groups in the following series of strata 'suddenly' extend themselves and thereby split up into numerous families, genera, and species. As a set-off other forms often contemporaneously die out.

§ 2. *Inter-relation between the greater systematic groups (families, classes, and partly orders).*

If therefore we concede that, on the whole, the higher forms chronologically follow the lower, do they originate therefrom?

(1) The Invertebrates appear together in the Cambrian formation, but clearly separated into all the families and most of the classes [1] which exist at present (see above, page 23). Meanwhile we are absolutely compelled to regard them all as originally separated groups.

'All the important phylae (families), sharply defined, reach back far into the Cambrian formation, and of those periods in which they might have been united we have no records.' [2]

We have known for a long time that the majority of the great groups of invertebrate animals are already quite distinctly separated in the Cambrian era.[3]

We cannot therefore deduce as originating from each other the classes of Invertebrates as they have been preserved to the present.

[1] Families, for examples, are the Worms, the Cœlenterata (corals, bryozoa, medusæ), in the Sea Urchins (Echinodermata).

[2] E. Koken: *Paläontologie und Descendenzlehre*, p. 12.

[3] Depéret-Wegner: *Die Umbildung der Tierwelt*, p. 233.

(2) Of the Vertebrates, the Fishes (Silurian) show no connection with lower forms: they appear as suddenly existent.

'We recognize the Fishes as the oldest Vertebrates, which already in the lower Silurian formation appear as clearly separated from the Invertebrates.'[1] They are, indeed, not only different from the Invertebrates from the beginning, but in the group of the primary fishes itself '. . . there are numerous quite different types co-existing but separate from the beginning.'[2] The same remark applies to the first Amphibia and Reptilia.[3] It is true that in the Permian system (in the Carboniferous there appear a few representatives) a peculiar group of animals widely prevailed—the so-called Stegocephalæ—which possess many characters of the present Amphibians (free larval condition and two occipital swellings) and others of the present Reptilia (scaly covering) in combination. In their appearance they resemble, for instance, salamanders (Amphibia) or lizards, crocodiles, and snakes (Reptilia).

But, contemporaneously, there already lived true reptiles (among them *Hatteria* up to the present time) and true Amphibia even previously (see p. 28). The Stegocephalæ cannot therefore be regarded as the common ancestors of the Reptilia and Amphibia.

With regard to the origin of the present reptilian

[1] Steinmann: *Die Geologischen Grundlagen der Abstammungslehre*, p. 203.
[2] *Ibid*. p. 206.
[3] *Ibid*. The origin of the Quadrupeds is still not cleared up.—Koken: *Paläontologie und Descendenzlehre*, p. 241.

orders (Turtles, Crocodiles, Snakes, and Lizards) nothing is known.[1]

The first birds, Archæopteryx, have toothed beaks, the vertebræ of the long tail remain separated, the free digits of the front limb (= hand or wing) carry claws. These are characters which occur now in reptiles. This bird has, furthermore, many other peculiarities confined to itself.

From this it was concluded that birds were descended from reptiles. But all attempts failed to trace a connection with any particular reptile.[2] The nearest birds (Cretacean) can by their habit be assigned to quite definite bird families—Laornis, for example, to the Geese—so that Steinmann can say 'each of the three well-recognized types of Cretaceous birds represents a separate ancestry.'[3]

The Mammalia show from the beginning 'two groups of lower mammalia quite clearly differentiated.'[4] Then they almost disappear during the immense Mesozoic period (= Secondary formation group), and appear again at the commencement of the Tertiary (Eocene)

[1] J. Bumüller: *Die Entwicklungstheorie und der Mensch*, p. 23 (according to Zittel's investigations).

[2] Steinmann: *Die Geologischen Grundlagen der Abstammungslehre*, p. 222: 'From the older strata of the Jura and from the Trias (which come next into consideration) we know practically nothing of small, long-legged reptiles of such a habit as the Archæopteryx demands as ancestor.' Depéret-Wegner: *Die Umbildung der Tierwelt*, p. 231: 'The Archæopteryx is, however, a true bird in its entire construction, and possesses beyond doubt a long line of ancestors which at present eludes our knowledge.'

[3] Steinmann: *Die Geologischen Grundlagen der Abstammungslehre*, p. 225.

[4] Depéret-Wegner: *Die Umbildung der Tierwelt*, p. 281.

period 'almost as fully typified and as sharply defined as to-day, particularly also such as were of unusual size or of peculiar travelling powers or habits of life, such as Cetaceæ (whales) Sirens (sea-cows) Bats, etc.'[1]

Summarizing, J. Bumüller says[2] '. . . We have therefore the remarkable fact that the placental Mammalia, which appeared first in the Tertiary period (see remark above, p. 29), is already split into all the ten orders in the oldest section of that period, viz. the Eocene. . . .' Where are the predecessors of these orders? Where are the transitional forms between them and the Marsupials which were there already in the Trias?

Deduction from § 2: Of a process of separation of the families and classes of the Invertebrates from each other, the higher from the lower, we know nothing since they appear contemporaneously as sharply separated in the Cambrian formation.

That the higher classes, and even many orders of the Vertebrates, have been evolved from the lower ones is, according to the actual results of investigation, in no single case other than probable. A single apparent exception here is that of the Birds. Archæopteryx was obviously a bird: the entire construction of the skeleton, the so characteristic form of the front and rear limbs (wings and legs), the possession of feathers, of which no reptile shows the slightest trace, separates

[1] Steinmann: *Die Geologischen Grundlagen der Abstammungslehre*, p. 233.

[2] J. Bumüller: *Die Entwicklungstheorie und der Mensch*, p. 76.

it entirely from all other classes. That the beak contained teeth while the present birds no longer possess them, that the caudal vertebræ had not yet become united as is now the case, shows that it was a different bird from the present ones, but otherwise demonstrates nothing. We are, however, accustomed to find in the first representatives of any type divergent and (by comparison with the present ones) curious characteristics.

It must therefore be assumed that the Birds bore teeth for a long period and only gradually lost them. In that case we should have here a so-called apparent regression (see p. 44).

§ 3. *Some palæontological 'laws' according to which the transformation proceeded within defined (narrower) groups (families, genera).*[1]

(1) *The law of increase of size.*

In the previous matter we have repeatedly called attention to the fact that the first representatives of a newly appearing group are often small and insignificant individuals compared with the later and sometimes gigantic forms within the same group. This is observed 'almost invariably in all classes of the animal world, but we find more numerous and clearer applications of the law in the group of Vertebrates than in that of the Invertebrates.' By careful investigation Neumayr, Waagen, Mojsisovics, and Hyatt have deter-

[1] We rely in this section particularly upon Depéret-Wegner's *Die Umbildung der Tierwelt*, chaps. ix. and x., where the whole literature of the subject is dealt with.

mined such evolutionary series[1] as, for instance, with Foraminifera, Sea Urchins, Brachiopoda, Ammonites, Nautilus among the Invertebrates, many fish groups (shark, lungfish), with Amphibia, Reptilia, and, before all, also for groups of Mammalia. 'Among the Mammalia the law of increase of size is demonstrated with the utmost possible clearness, so that for the modern palæontologist it may be used as a veritable touchstone in connection with the reconstruction of genealogical trees.' In many cases the entire 'evolution' of the offspring is confined to increase in size: the organic characters remain almost unchanged.[2] The 'law' is, however, not general; for instance, it does not apply at all to insects (see p. 38).

(2) *The law of specialization and differentiation within more defined (narrower) groups.*

We have already several times stated that the representatives of an organic class on its first appearance show simple forms which are not yet 'specialized.' Frequently such groups split up later into numerous new forms, species, genera, and families. It has now been observed that this rich development arises through the original individuals changing in quite definite directions: for instance, in the Cephalopods the shell

[1] Depéret-Wegner: *Die Umbildung der Tierwelt*, p. 181.

[2] A classic example is Brachyodus (Depéret-Wegner: *Die Umbildung der Tierwelt*, p. 185), which is increased from the size of a hare to that of a rhinoceros, and yet it retains the generic characters perfectly and only forms another 'species' than its dwarf ancestors. Also in the hypothetical evolutional history of the Horse, the increase of size plays an important rôle; it is true that in this case it is accompanied by other important modifications.

tends to become more and more convoluted, more and more decorated with lines and excrescences, or some definite organ is added and thereby indirectly the entire organism becomes more and more specialized for a definite function (mode of progression, transformation of the limbs for swimming, flying, running, digging, etc.), and as this proceeds always in the same direction it becomes more and more adapted to the particular function concerned. With these phenomena is associated the well-known example of the so-called horse-foot series, which demonstrates how, from a normal five-toed foot, the one-toed foot of the present horse has been quite gradually evolved, as a constantly one-sided, and therefore, in a certain sense, a more and more perfect, adaptation for speedy running. With it is furthermore associated the evolution of the paddle hand of the Sirens (sea-cows), the evolution of defensive and gripping weapons—for instance, the horns of the stag which from small beginnings arrived at colossal dimensions in the extinct gigantic deer; also the tusks of the Proboscidæ, etc.[1]

'Specialization' signifies, therefore, the development in one direction of an organ or of the entire organism. What causes are effective are not always demonstrable; probably the necessity of purposeful adaptation to changed environments was the reason.

'Differentiation' is the development of numerous variations of one and the same fundamental type by specialization of separate individuals in different direc-

[1] Depéret-Wegner: *Die Umbildung der Tierwelt*, chap. xx.

tions. As fundamental types we regard those of the first individuals which have become known to us. That is perfectly correct if at least the first representatives are few and all similar as between themselves, and if it can be accepted that precisely those individuals formed, in fact, the starting-point of later forms (species and genera). In that case it is clear that one type, originally confined to one or a few species, becomes varied or differentiated.

With the numerical increase of the individuals both animals and plants naturally incur the necessity of dividing their resorts or habitats, they occupy different elevations in valleys or on the mountains, deeper or shallower water, and more humid or drier climates, etc. This gives the impulse to varied specialization. The type itself becomes varied by the varied specialization of the separated individuals.

The phylogeny of the Insects may be studied more in detail to elucidate the above. For this group we possess A. Handlirsch's great work, 'Die fossilen Insecten und die Phylogenie der rezenten Formen,'[1] a comprehensive presentation and consideration of the whole of the discoveries so far made.

The oldest insects, the Palæodictyoptera (Fig. 5), appear in the lower strata of the very productive Carboniferous system. They were of considerable size, many of them as long as the hand or even the arm, with six legs and four or six wings. The vein system

[1] Vol. II, Leipzig, 1906–1908. An exposition by Handlirsch himself appeared in *Die Umschau*, 1909, p. 588.

of the wings was well developed; the antennæ consisted of numerous segments; the rear segments of the body frequently bore gill-like appendages.

By the appearance of their masticating and well-developed jaws they were carnivorous. Their larvæ lived probably in water like the present Ephemeridæ. The body was equally segmented (homonomous).

The whole of the primary insects form together the one order of Palæodictyoptera, but this consisted of twenty-two families and 115 species. The numerous orders of the insects of the present day are still absent. In the upper strata of the Palæozoic period insects appear which 'undoubtedly present a definite tendency towards the modern insect orders.'[1] Handlirsch considers them as 'transitional groups between the primary groups and those to-day existent, the Orthoptera, Woodlice, Dragon-flies, Ephemeridæ, Hemiptera, etc.' In the old terrestrial formations, however, we find already true Woodlice, Ephemeridæ, and Locusts: 'the Palæozoic fauna is therefore totally different from the modern and much more uniform'[2] (Fig. 6).

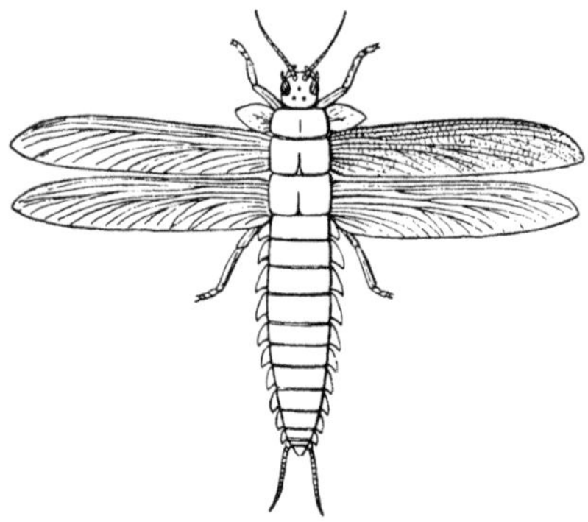

FIG. 5.—RECONSTRUCTION OF THE PRIMARY INSECT (reduced). (*After Handlirsch.*)

[1] *Die Umschau*, p. 589. [2] *Ibid.*

Our illustration shows such a transitional group from the Palæozoic system which Handlirsch regards as the ancestral form of the Scorpion flies, Phryganidæ, Flies, and Butterflies.

In the Mesozoic system the primary form is no longer found and the transitional forms are scarcer. 'Nearly all the insects found in these formations can be allocated without difficulty to the now existent orders, although they differ sufficiently from many of the present forms to be considered separate families or at least genera.'

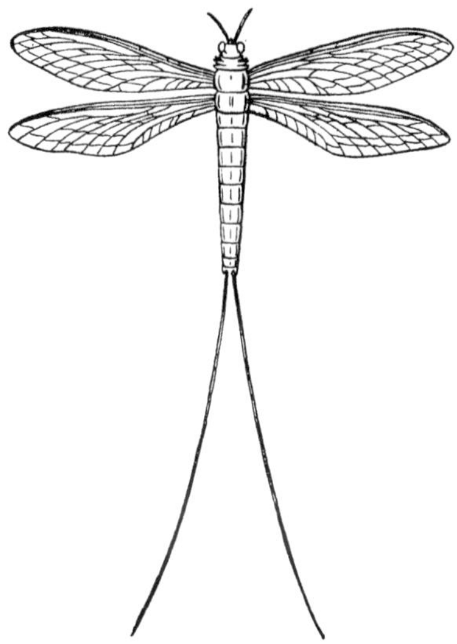

FIG. 6.—A TRANSITIONAL FORM FROM THE CARBONIFEROUS SYSTEM. (*After Handlirsch.*)

There appear true Locusts, Grasshoppers, Stick Insects, Beetles, Phryganidæ, Dragonflies, Hemiptera, Butterflies, Bugs, etc., so that at the end of the Jura period (Figs. 7–9) all the chief groups of the insect world are existent with the exception of the Earwigs, Termites, Woodlice, Field-cricket, and the true Lice (infesting warm-blooded animals), Fur-eaters and Fleas (Parasites). Within the orders, however, there are still lacking many now widely extended families, such as Wasps, Ants, Bees, true Flies, Gall-wasps, etc.

The numerical relations differ also from those of to-day.

That which is lacking in the Jura 'we find already richly represented in the Tertiary deposits. All families now agree, as do very many genera and even several species, with those now existent, so that the difference consists now almost entirely in the numerical relations and in the geographical distribution.'[1]

According to the above the whole of the present

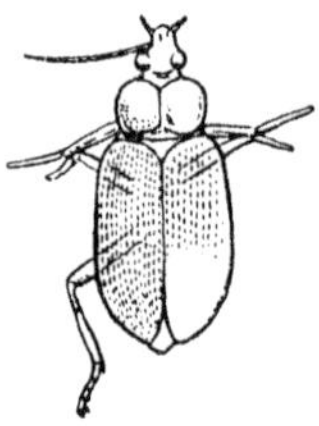

FIG. 7.—BEETLE (Jura).

FIG. 8.—PRIMARY WOOD WASP (Jura Hymenoptera).

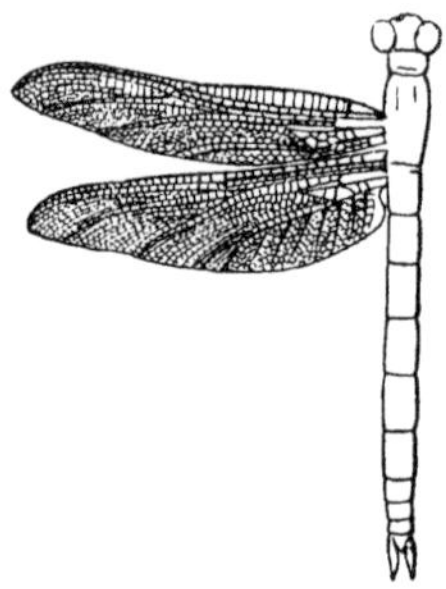

FIG. 9.—DRAGON-FLY (Jura).

FIGS. 7–9. (*After Handlirsch.*)

insect orders with their subdivisions have arisen by differentiation of one original type. This occurred through changes in the originally similarly segmented body—for instance, in the Beetles and Hymenoptera, into three clearly divided main sections: head, breast, and rear segments. Other recent offshoots, like the Termites, show a still more uniform segmentation of the whole body. In the Thysanura this is complete.

The formation of the antennæ, also the number of tarsi and the venation of the wings afford opportunities

[1] *Die Umschau*, p. 590.

enough for specialization and particularly also the formation of the jaws, which are already differentiated for masticating, stinging, sucking, etc.

In many cases the specialization was associated with regression, with reduction of the number of wings and segments.

As an external impulse to variation Handlirsch cites the Permian glacial epoch which, particularly, as a result of the general cooling, may have caused the transition of the larval forms into perfectly metamorphosed ones as 'being generally better fitted for cold seasons.' For the late appearance (and origin) of the Wasps, Ants, Bees, and Gall-flies, Handlirsch considers that in the appearance and rich development of the flowering plants, a sufficient reason is to be found.

Wasmann[1] expresses his opinion regarding the history of the origin of insects, as Handlirsch presents it, as follows: 'In the hypothetical history of insect origin certainly a manifold differentiation and specialization of the insect type has taken place, which was connected, partly with an increased development of certain characters, and partly with a reduction in others. Of an enhancing of the entire organic type naught can be said. . . . A wingless worker ant can by no means be considered as "a more perfect insect" than a six-winged Palæodictyopteron was. The primary insects and those of their present descendants which, like the Ephemeridæ, have retained many of the

[1] *Die progressive Artbildung und die Dinarda-Formen in Natur und Offenb.*, 1909, no. 6, p. 333.

original characters, may well be considered as the most primitive insects, but not as the lowest.'

(3) *The law of the limitation of such specializations in one direction—Explanation of the extinction of such forms.*

(*a*) The almost sudden disappearance of many multifarious groups has long presented a particularly attractive problem. In the Primary period there appeared, for example, Trilobites in the Cambrian system, they passed their prime in the Silurian and Devonian, and entirely disappeared in the Permian. The same thing is observed in the Primary epoch with the Graptolites, Cystoidæ and Blastoidæ (sea urchins), with the Tetracorals and Euryptidæ (gigantic crustaceans).

In the Secondary period there may be similarly observed the appearance and disappearance of the Belemnites (Thunderbolts), Hippurites, and the gigantic Saurians (Ichthyosaurus, Plesiosaurus, and Pterosaurus).

In the Tertiary period occurs the appearance and entire extinction of many large groups of Mammalia.

(*b*) It has been endeavoured to account for this enigmatical disappearance of entire classes and orders of animals by catastrophes (Cuvier), or also by epidemics and starvation (Neumayer, Quenstedt).

But the two 'laws' already discussed afford quite another explanation.

It had always been noted that 'the species of a group find themselves on the eve of disappearance precisely when they have attained the maximum of prosperity, either with regard to the dimensions of their bodies or in the perfection of their weapons of attack and defence.'[1]

In the end, however, horns of two metres in breadth, as were borne by the gigantic deer, must become directly detrimental; reduction of a specialization carried so far appears, however, to be impossible (law of the irreversibility of evolution). The increase in size has also its limits; if overstepped the size leads to clumsiness and unwieldiness. If the environment then be altered, such forms, so peculiarly modified in the one suitable direction, are doomed to extinction.[2]

Such environmental changes on a large scale certainly occurred. We may consider only the various mighty ice invasions which repeatedly took place: the first certainly occurred already in the pre-Cambrian epoch,[3] another in the Permian,[4] and the last great one was that of the Diluvium. We may further consider the frequent incursions of the ocean, and the climatal oscillations, which resulted in our finding in one and the same region the remains of tropical or sub-tropical

1 Depéret-Wegner: *Die Umbildung der Tierwelt*, p. 219.

2 Recently the palæontologist R. Hörner has exhaustively discussed the question of Extinction in his work *Das Austerben der Arten und Gattungen sowie der grosseren Gruppen der Tier- und Pflanzenwelt*, Graz, 1911. In essentials he agrees with Depéret's views.

3 J. Walther: *Ueber algonkische Sedimente*, in *Naturw. Rundschau*, 1910, p. 158. E. Kayser: *Lehrbuch der Geolog. Formationskunde*, p. 50, Supplement. (Clear traces are found of same in N. America, Norway, and China.)

4 E. Philippi: *Ueber die Permische Eiszeit*, *u. Zentralblatt fur Mineralogie, Geologie, und Paläontologie*, 1908, p. 353. The author regards it as 'proved.'

animals and those of the present Siberian or high alpine fauna superposed.

(4) *The phenomena of regression and of convergence (Law of Convergence).*

(*a*) Regressive Evidence.—Many animals experience during their individual lives, under the influence of particular conditions (parasitism for example, or transfer to an established mode of life), a clear depreciation of many organs through regression and reversion. The digestive apparatus can be entirely transformed—for instance, in the female of many parasitic crabs (Lernaiden),[1] as also that of locomotion, which in both cases are no longer purposeful for movement from place to place and therefore disappear (Lernaiden for instance) or serve other purposes (direction of nutriment towards the mouth) as will the species of Lepas, the so-called 'Duck Mussels' in the Crab group of Centipediæ.

Palæontology, in the opinion of serious investigators, has now afforded some evidence of the way in which, in the course of geological time, a reversion of some sections of organisms gradually occurred. Thus the presumed parents of the horse of to-day possessed fully-developed lateral toes which now appear only as 'sesamoid bones' and are hidden beneath the skin. It is furthermore accepted that the Birds, whose oldest representatives in the Jura and the Chalk formations all possessed teeth, lost them gradually. In young parrots and in the embryos of other birds of

[1] R. Hertwig: *Lehrbuch der Zoologie*, Jena, 1907, p. 382.

to-day there is still discoverable traces of tooth-bearing jaws.[1] Furthermore the so-called medial eye on the head, which was well developed in many palæontological reptiles, has retrograded and been lost. In *Hatteria*—this relic of a long-vanished world—the medial eye is still existent but is hidden under an opaque skin.[2]

These evidences of retrogression, which appear in the forms of the present day as so-called 'rudimentary organs,' have, in the opinion of palæontologists, contributed much to the transformation of the animal world. The alteration frequently extends only to a system of organs which, by reason of a transfer to a new mode of life, have become superfluous, and frequently only to a part of same; hence in the opinion of many investigators the so-called fins of the whale are a positive adaptation of the front limbs to swimming, and thereby no retrogression but a 'specialization,' while the hinder parts (legs and pelvis) have dwindled and now only remain as an insignificant bony rudiment buried within the fat of the whale.

This, however, brings us already to the so-called convergence evidence.

(*b*) Evidence of Convergence.—It is frequently observed that animals which, systematically, stand far apart, exhibit changes in the same direction and develop these further, so that by their further evolution they approach each other nearer than they were

[1] Depéret-Wegner: *Die Umbildung der Tierwelt*, p. 202.
[2] *Ibid.*

before: they converge towards each other. Thus, for instance, many mussels which by their construction belong to various species and genera, commence at the same time or successively to alter their shells in the same direction. Since now it is the shells alone which are preserved, in those cases where the remainder of the varied organism has left no trace we can no longer know with what species or genera we have to deal.

Yet the similarity thus engendered is mostly only superficial since it extends almost always only to the shells, scales, and epidermal plates, as we see in the Molluscs, Fishes, and Reptiles; or, since only some of the organs—for instance, those of locomotion—are similarly formed, the peculiarities of the whole organization are never perfectly eliminated. It is, therefore, by convergence that we explain how, within quite different groups of mammalia, a most deceptive similarity of the jaw construction is observed; in this case this is undoubtedly caused by habituation to the same kind of nutrition.[1]

By convergence is explained also the entire 'fish resemblance' of the Whales, which, when habituating themselves to life in the water, were guided, so to speak, by the water inhabitants *par excellence*, the Fishes—that is, they changed their forelimbs (arms) into

[1] If only separated plates (shields) or separate teeth are found then we cannot tell to which animal group they should be assigned. Hence recently a polygonal epidermal plate was ascribed by Filhol to an extinct armadillo (Mammalia), but later an almost exact replica was found, but this time on the head of a reptile (Placosaurus).—*Umbildung der Tierwelt*, p. 208.

fins (paddles) and their hind limbs became entirely useless. The tail became likewise transformed into a rudder, which, indeed, is the case with the Fishes, etc. Yet mammals they remain all the same, since the mammalian nature is not in itself contradictory to an aqueous existence ; but everything that was specially adapted for a land life must be transformed to fit the new water one and that only.

Hence they converged ever more and more in external features towards the Fishes, with whom otherwise they are not at all related.

Naturally such evidence of convergence must be established or rendered probable by exact comparison of changes in both the converging groups. Of the Whales there are lacking entirely fossil remains of any kind whatever.[1]

In point of fact so far no evidence of convergence has been palæontologically determined, by which even only the original generic characters have been perfectly eliminated.[2] 'On the whole I think,' says Depéret, 'that the evidences of convergence which were asserted in connection with nearly all animal groups were greatly overestimated. In the majority

[1] Steinmann (*Die Geologischen Grundlagen*, etc., p. 235) will not accept the idea of descent of the Whales from quadrupedal ancestors. He endeavours to explain their descent from definite groups of the great Mesozoic Saurians. There is much to be said for his opinion, but 'very convincing' it is not.

[2] Depéret-Wegner: *Die Umbildung der Tierwelt*, p. 205. Professor Fleischmann (*Die Darwinische Theorie*, Leipzig, 1903, p. 263) writes, regarding the origin of the whale, a veritable satire, most delightful to read, but which avoids all explanation.

of cases the resemblances of this kind are very superficial and can be easily explained by the process of adaptation to functions common to both.' [1]

Conclusions derived from palæontology:

We have seen that the changes or transformations which are experienced by the organisms never exceed the limits of the families and classes, nor generally those of the orders (§ 2).

According, therefore, to the present position of science, there is no unlimited transformation in the animal world.

It is true that with the Vertebrates, on the whole, the higher classes appear after the lower (§ 1). To show how both, viz. the appearance after each other and yet no derivation from each other, can be brought into accord is reserved for the evolutionary hypotheses. We will express a supposition relative thereto later on.

In § 3 we saw in what way originally like or similar forms could arrive at varied appearances. This happens through an ever-increased adaptation to quite definite modes of life by which, in some cases, the whole organism is transformed—generally, however, only separate organs. There is often observed also an increase of size, in conformity with a recognized law, mostly in conjunction with other changes, but in a few there is increase of size alone.

It may also happen that animals originally widely separated in kind become more similar by variation in the same direction.

[1] Depéret-Wegner: *Die Umbildung der Tierwelt*, p. 213.

It therefore happens that the appearance of the organisms is constantly changing, and forms appear which are new and specialized, i.e. adapted to quite definite modes of existence, nutrition, habitats, etc. But in no case does the entire change, which the organisms finally show in comparison with the original form, go so far that offspring and ancestors can no longer be united within the same systematic class. Generally the totality of the descendants and ancestors, despite all 'evolution,' still form the same order, the same family, and sometimes even the same genus. 'Genealogical trees' or pedigrees which ignore all systematic classification are simply illusory.

The 'evolutionary series' (pedigrees) of some organisms, the course of which may be followed more or less without hiatus, never present such considerable deviations from each other that the new forms (species and genera) must be arranged 'in separate natural families.'[1]

B.—Results of Palæobotany (Evolution of the Plant World).

With the history of the evolution of the plant world we are less acquainted than with that of the animals. This is due to various causes.

In order that organisms may be preserved in a fossil form they must be withdrawn from the corroding influence of the atmosphere, particularly the free access of oxygen. This happens generally, and is best effected, by deposition of mud, sand, or even of other

[1] Depéret-Wegner: *Die Umbildung der Tierwelt*, p. 250.

organisms under water. For this reason land dwellers rarely become fossilized.

For the investigation of the old-world fauna this circumstance is less serious than for the flora, since the greater number of the animals are and were inhabitants of the water and particularly of the ocean. But of the higher plants[1] which inhabited the sea we know nothing.

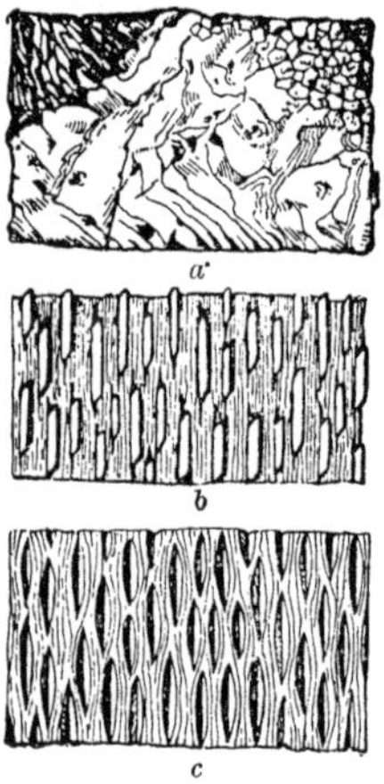

Fig. 10.—Sections of Beech.
a, with bark; *b*, with bark removed showing bast; *c*, with bast removed to true wood. (*After Gothan.*)

Hence it happens that formations which are exclusively marine and, particularly for our present inquiry, very important stratifications—the pre-Cambrian, Cambrian, Silurian, Mussel Chalk, and others—only now and then offer examples of submerged land plants. With regard to the flora we gather exact data only from those periods and regions where extensive areas of land were marshy or, owing to great aerial humidity, were covered with forests. Only under such circumstances were the conditions existent for such processes as led to the formation of the coal seams, in which we find entire generations of successive growths,[2] and these in the best condition. It must also be considered that none of the primeval plants have

[1] 'Higher plants' we only accept temporarily in the sense of their higher systematic position, as they are recognized generally by botanists.

[2] H. Potonié: *Die Entstehung der Steinkohle und der Kaustobiolithe überhaupt*, Berlin, 1910.

practically survived as perfect examples or in the form of larger connected fragments, but mostly are found as separated portions of one and the same plant at different places—for instance the stem here, the leaves yonder, and the seeds and fruit somewhere else.

It is clear that under these circumstances it is only in exceptional cases that the properly associated parts can be recognized as such.[1]

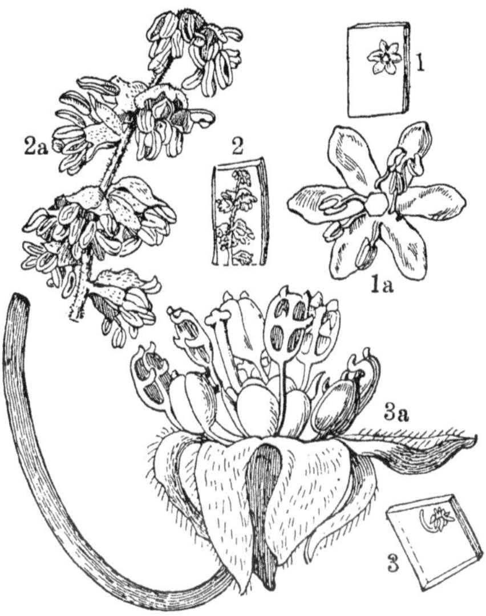

FIG. 11.—REMAINS OF PLANTS IN SAMLAND AMBER.

1. Sambucus (Elder) Flower, $\frac{3}{4}$ nat. size.
1a. The same magnified.
2. Portion of a male Oak Catkin.
2a. The same magnified.
3. Cinnamon Flower.
3a. The same magnified.

(*After Gothan.*)

To this must be added that one and the same plant occurs in quite different states of preservation, with or without bark (Fig. 10), stone kernels, pith tubes, etc., so that quite different forms of growth may present themselves and receive also different names. Fig. 10 will help to a comprehension of this. It shows us three

[1] Dr. W. Gothan: *Entwicklung der Pflanzenwelt, Osterwieck am Harz,* 1909, p. 6. This little volume of the collection *Die Natur* is much to be recommended. We shall follow it generally in our arguments. Gothan obviously relies greatly on Potonié.

conditions which may be easily noted in the same decayed stem. In all these conditions we find fossil trees. Flowers are really only known to us by enclosure in amber (fossil resin) (Fig. 11).

Nevertheless it is possible, though certainly only on rough lines, to conceive an idea of the succession of series and the connection of the greater groups, of the process of transformation within separate types and in some degree also of the reasons and causes of same.

§ 1. *Brief purview of the chronological succession of the larger plant groups.*

Gothan adopts, for the history of plant life, a somewhat different limitation of the three chief periods, quite logically according to the principles which we have treated more in detail above.[1]

According to this the Palæozoic era closes with the lower Permian, the Mesozoic with the lower Chalk where the Cænozoic commences.

(1) *Oldest discoveries of plants.*

According to Gothan[2] the graphite masses which are found even in the Primary rocks (gneiss and Primary slate) point with certainty to the existence of organic growths. Although one might be inclined, owing to the dependence of the animals upon plants, to deduce

[1] The first appearance of new types, high development, and predominance of previously sparsely represented ones or disappearance of other previously most diversified forms. (See above, p. 24.)

[2] Dr. W. Gothan: *Entwicklung der Pflanzenwelt, Osterwieck am Harz,* 1909, p. 15.

at least a contemporaneous origin for both, yet the occurrence of graphite does not suffice to afford a proof of this, 'because,' as Kayser says,[1] 'in no case should the accidental chalk and graphite deposits of the gneiss be regarded as proofs of organic life in the Primary period, since chalk and graphite, as it can be proved, may arise also on inorganic lines.'

The first certain traces of growth in recognizable remains we find in the Silurian formation. The alleged seaweeds of the pre-Cambrian and Cambrian systems must to a large extent be otherwise explained.

After the experiments reported by Nathorsh it is impossible longer to refrain from the opinion that 'a large number of the smaller fossil algæ are either the results of processes in rock formation or animal tracks, or furthermore are produced by running water or plants moved by water,' or, as is stated farther on, 'the remains of tissues of more highly organized plants.'[2]

The Silurian remains, fern fronds, and large masses of algæ (bladder algæ—Siphonæ); show us that there were already representatives of the higher systematic groups. By reason of the simple arrangement of the veinlets[3] in the fronds the first ferns are designated as 'primitive.'[4]

[1] *Formationskunde*, p. 21. The evidence was mainly provided by Weinschenk.

[2] A. Schenk, in Zittel's *Handbuch V*, p. 233. By this the existence of vascular plants was indicated. More details cannot, however, be recognized owing to the great decomposition of the materials.

[3] The veinlets are bundles of conducting vessels by which water and earthy salts in solution are carried to the assimilating tissues.

[4] More details are given in § 3, where we shall treat of the differentiation of the fern type.

To all appearance there also already existed in the Silurian era the predecessors of the Sigillaria (which developed later so richly in the Carboniferous) (Figs. 12 and 13) in the form of Bothrodendræ, if the age of the localities of the finds (Harz) be correctly estimated.

Even Gymnosperms are found already in the group of Cordaitæ[1] which also play a very prominent rôle in the Carboniferous system.

(2) *The further development of the flora.*

In the Devonian formation the discoveries of land or rather marsh plants become already more frequent. With the ferns there appears a further group of the highest Cryptogams, the Protocalamariaceæ; these are the Equisetum trees—as yet but few in number and fairly alike among themselves.

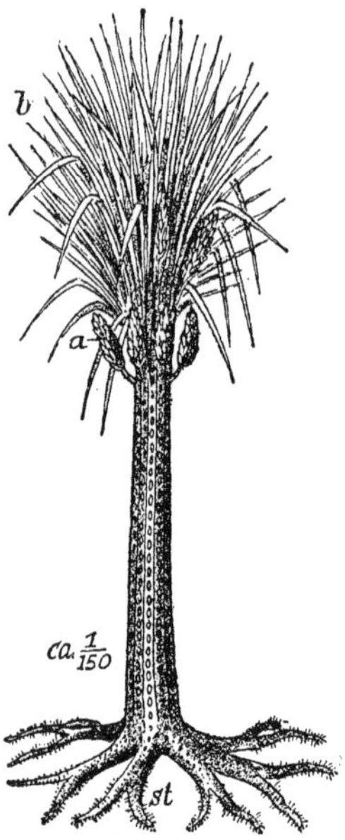

FIG. 12.—RECONSTRUCTED SIGILLARIA.

a. Flower. *b.* Leaves.
st. Rootstock (stigmaria).

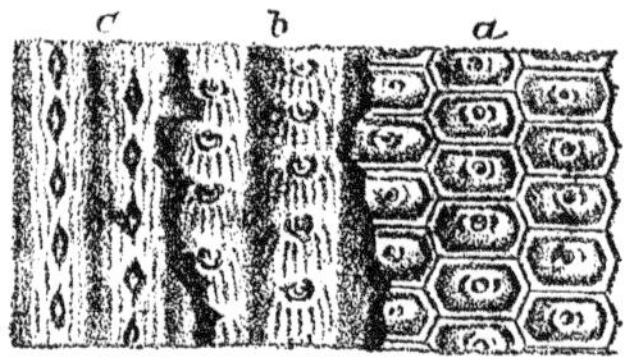

FIG. 13.—PORTION OF TRUNK OF SIGILLARIA SHOWING VARIOUS STATES OF PRESERVATION.

a. Bark entire.
b and *c.* Bark removed.

[1] Zittel's *Handbuch V*, p. 251.

By the evolution of numerous variations this primary type attains its period of maximum development also in the Coal period.

The Cordaites become more numerous; they also begin to vary in form.

In the Devonian of Bohemia Potonié has found also the remains of Ginkgo-like plants[1] as representative of a further group of Gymnosperms.

In the coal seams of the Carboniferous age, according to an appropriate remark of Potonié,[2] tropical marshlands (*Sumpfflachmoore*) have come down to us in a fossilized state, and by both these terms—'tropical' and 'marshland'—the flora of that period appears to be well indicated.

The growths which, in conjunction with flowering plants, also at present form the main constituents of such 'reedbeds,' are represented, and in a manner befitting the most luxuriant environmental conditions, by gigantic tree-like forms of the three classes of Pteridophytes, the true Ferns; then the Club Mosses

[1] J. P. Lotsy: *Vorlesungen über Descendenztheorien* (*Mit besonderer Berücksichtigung der Botanischen Seite der Frage*) *II*, Jena, 1908, p. 466. The Ginkgos externally resemble our leafy trees. The only species still existent—*Ginkgo biloba* (on account of the two-lobed leaves)—is indigenous in China and Japan, but as quite solitary specimens. It may frequently be seen in our parks. They form a quite peculiar group, which Lotsy thus describes: 'Gymnosperms with conifer-like wood with male and female inflorescence widely differing from the Cycads, but with ovaries and seed resembling those of Cycads.'—Lotsy: *Vorträge über Bot. Stammesgeschichte*, II, Jena, 1909, p. 778. (Not to be confounded with the work above cited.) (Cycads and conifers are two chief groups of recent Gymnosperms.) The Ginkgo was therefore no more a transitional form in the Carboniferous period than it is to-day, as we shall see later on.

[2] *Die Entstehung der Steinkohle*, etc., p. 186.

(Lepidodendron) (Fig. 14), and Sigillaria[1] and the Calamites.

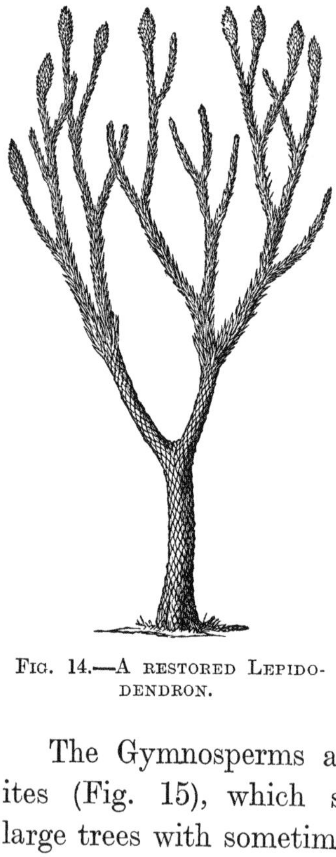

FIG. 14.—A RESTORED LEPIDODENDRON.

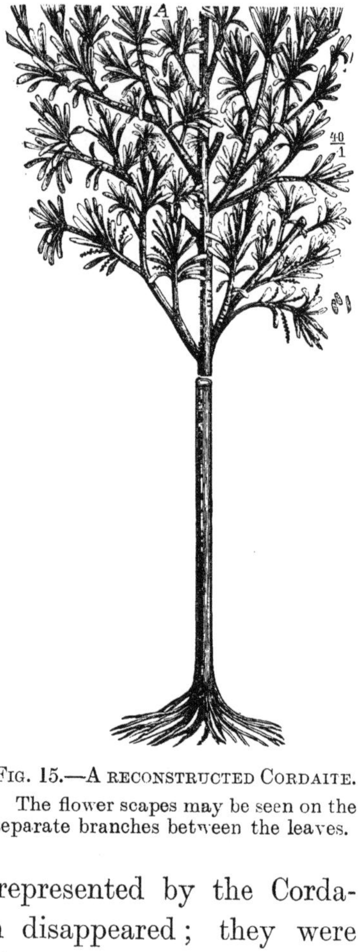

FIG. 15.—A RECONSTRUCTED CORDAITE. The flower scapes may be seen on the separate branches between the leaves.

The Gymnosperms are represented by the Cordaites (Fig. 15), which soon disappeared; they were large trees with sometimes gigantic leaves, which there-

[1] Schuppenbäume (lit. scale trees: Lepidodendræ) derive their name from the cushion-like elevations (scales) on the bark which bear the scars of the fallen leaves. The Sigillaria (lit. seal trees) do not show these cushions or scales; the leaf scars lie flat upon the bark, and are hexagonal.

fore do not agree at all with the needles and various foliage of our Gymnosperms but are rather remindful of the parallel-veined ones of the Monocotelydons (e.g. the Liliaceæ).

FIG. 16.—RECONSTRUCTION OF A TRUNK OF LYGINODENDRON OLDHAMIANUM, A FERN-LIKE GYMNOSPERM. (*After Scott.*)

The three smaller branches seen at the top of the figure with contracted apparent leaves are regarded as the inflorescence, with the empty seed-vessels at the ends of the stalks. (*After Oliver.*)

Until recently it was a standing expression in all text-books that the Palæozoic age, and particularly that of the Coal period, formed the epoch of fern growth, i.e. of seedless plants. Recently, however, a whole series of families which until then, owing to the form of their foliage, had been regarded as true ferns, have had their systematic classification entirely upset. It was found, for instance, that a number of isolated seeds could, with a probability bordering on certainty, be associated with 'fern' leaves and 'fern' stems.[1]

FIG. 17.—The seed (L. Oldhamianum, Fig. 16) is enclosed in a husk which is covered with glands. (*After Biol. Zentralblatt.*)

Oliver and Scott particularly and Stur previously have done meritorious service in the determination of this most important discovery.

Figs. 16 and 17 show a 'fern' stem and the seed belonging thereto, which in exterior form resembles a hazel nut. The new group of these Carboniferous Gymnosperms received the provisional names of 'Pteridospermen,' i.e. seed-plants with fern-like foliage. They are not intermediate forms since there are to-day Gymnosperms with fern-like leaves, e.g. the sago or fern palms.

We observe then, in the history of the plant world, that with the progress of palæontology the systematic

[1] Compare with this the instructive statement by F. W. Oliver: *Ueber die neuentdeckten Samen der Steinkohlenform*, in the *Biol. Zentralblatt*, 1905, p. 401; Lotsy: *Vorträge über Bot. Stammesg.*, II, p. 706; Potonié, in the collected work, *Die Natürlichen Pflanzenfamilien*, published by Engler and Prantl: I, Part IV, p. 780.

groups of higher standing must be referred to an older and older period. 'It is not long ago,' says Lotsy, 'that it was thought that the seed-bearing plants were of comparatively recent origin and that at least in the Coal period they were entirely absent.' Now the Cryptogams (non-seed-bearing plants) are not even conceded predominance in the later Palæozoic period (Fig. 18). 'Gradually the Ferns, one after the other, showed themselves to be seed-bearers, and it is difficult to say to what number relatively this alteration will extend. It will probably be a large one.' [1]

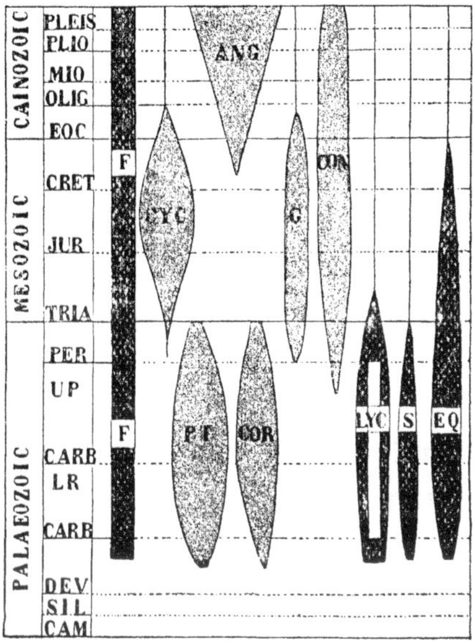

FIG. 18.—ANG, Angiospermæ; CON, Coniferæ; COR, Cordaiteæ; CYC, Cycadophyta; EQ, Equisetineæ; F, Filicineæ; G, Ginkgoales; LYC, Lycopodineæ; PT, Pteridospermæ or Cycadofilices; S, Sphenophyllales.

It is seen that the Palæozoic seed-bearing plants—PT and COR—approximately equal the non-seed-bearing, F, LYC, S, and EQ. (*After Oliver.*)

Yet must we agree with Gothan's warning against over-hasty conclusions.[2] The true fern nature of a leaf is certainly only beyond all doubt when we find the spore heaps on the under side of the foliage or elsewhere, since thereby they are directly shown to be spore-bearing

[1] *Biol. Zentralblatt*, 1905, p. 414. [2] *Entwicklung der Pflanzenwelt*, p. 35.

plants (not seed-bearing). But even the constant absence of spore-heaps by itself is no proof that the leaves belong to a seed-bearing plant, since ferns can propagate themselves purely vegetatively by stolons and offsets.

Lepidodendrons, Sigillaria, and Cordaites die out in the Old Red Sandstone—that is in the formation immediately succeeding the Carboniferous.[1] As separate individuals and species here for the first time appear clear remains of Ginkgo trees, also Conifers (Walchia) which are related to the Araucarias [2] and Cycads.

In the Permian limestone the development of the new forms commences vigorously, the gigantic Pteridophytes and the Cordaites have disappeared, so that there is a sufficient reason for beginning, with the Permian limestone, a new period—the Mesozoic—which shall be 'the era of the Gymnosperms.'

Angiosperms, i.e. flowering plants with covered seeds, which, as Monocotyledons and Dicotyledons, form the great bulk of our present flora, are absent in the Mesozoic period. The groups of Gymnosperms develop more and more the forms which approximate nearest to our present ones. In the Chalk our species are already to be found.

In the lower Chalk there appear the first remains of flowering plants: therewith begins the Cænozoic

1 In the Red Sandstone we meet again with some traces. They appear then, however, to be entirely extinct, i.e. leaving no successors even of different appearance.

2 Araucarias are represented by the so-called 'monkey puzzles' which are frequently cultivated.

period. 'These new plants, almost from the very commencement, appear in such quantities that the Gymnosperms, which earlier predominated, to say nothing at all of the Ferns, etc., had to retreat to the background.' [1]

The classification of the separate families depends upon the leaves, which in most cases naturally cannot secure absolute certainty. Stems are rare and the flowers we know really only by enclosures in amber. The Catkin-bearers and the Laurel family are in any case very old groups. Since the Oligocene period we find, in ever greater numbers, species similar to or quite like the present ones, and often quite 'specialized.' In the Pliocene there already lived, for example, the Silver Poplar, the Aspen, the Red Beech, the Mountain Maple, etc.

Conclusions from § 1.

(1) The earliest history of the plant world is so far entirely unknown to us; we know neither when the first growths appeared, nor how they looked.

(2) It must be accepted that already, at the time when the Cambrian and pre-Cambrian formations were deposited and the animal world was already so grandly developed, a rich flora of some kind also existed, since the animal world is dependent upon the plant world.

(3) Despite the great imperfection of the evidence obtained, and the fact that it is only remains of

[1] Gothan: *Entwicklung der Pflanzenwelt*, p. 86.

inhabitants of damp and humid habitats that have been preserved in great numbers which do not permit of an absolutely certain decision, it is very probable that the Gymnosperms followed the Pteridophytes and the Angiosperms the Gymnosperms.[1]

§ 2. *Inter-relations between the larger groups, families (series), and classes.*

As above for the animal world, we put the question also here, whether the larger groups were derived from each other in succession, so that, for instance, the Angiosperms represent only a higher development of the Gymnosperms.

(1) In the system generally in use, there stand upon the lowest step the so-called Thallophytes (Algæ and Fungi), then follow the Mosses, the Ferns (really ferns), Club Mosses, Equisetums, and Hydropterides (Water-Ferns), the Gymnosperms, and, finally, the Angiosperms (Monocotyledons and Dicotyledons).

Whence the Ferns, the first indubitable plant remains, come, no one knows. In the pre-Silurian formations we know certainly of no mosses which, purely *a priori*, come next to them in consideration, and not even of clear traces of Thallophytes which permit of any recognizable connection with ferns.

[1] By 'Ferns,' 'Gymnosperms,' and 'Angiosperms' we mean here only those foliage plants or trees, in short, those forms which the layman finds so named only in botanical works. Science differentiates all these plants by a double generation (of which more later). The sense of our above remark is simply that Ferns, Gymnosperms, etc., followed each other successively.

(2) The Gymnosperms are themselves very old forms: judging only by the palæontological remains, they belong to the oldest flora. Of a genetic connection with ferns or other plants there is naught to be said.

The new class of 'Pteridosperms' of the Carboniferous formation changes nothing in this respect, since despite the fern-like foliage they form no intermediate link between Ferns and Gymnosperms but, as seed-bearing plants, are pure Gymnosperms. Even to-day we have, among the Cycads, which are true Gymnosperms, forms with fern-like foliage; and one Cycad (*Stangeria paradoxa*) (a species now existent) ranked long as a fern, until its flowers were discovered.[1]

The determination of the systematic classification must, even according to Potonié,[2] who is a strong advocate of transitional forms, be made dependent upon the discovery of the seed or of the spore heaps. If seed be found then the foliage and stems appertaining thereto are those of true seed-bearing plants, in the other case they are true ferns. He himself grants that agreement in many anatomical characters—thickness of growth, venation, form of leaf—may all be attributed to 'adaptation to the same mode of existence.'

(3) Representatives of the true flowering plants or Angiosperms '. . . appear at first in the Chalk and in forms of such high organization as to agree

[1] Gothan, p. 35.
[2] Engler and Prantl: *Die Natürlichen Pflanzenfamilien*, I, Part IV, p. 789.

with the Dicotyledons of the present day. Precursors of these first Dicotyledons belonging to older formations are entirely unknown to us.'[1] We have already emphasized above the fact that the highest systematic group appeared at the outset as a numerous one and in genera and families which are still existent. With regard to the relations of the Angiosperms to the Gymnosperms Reinke says:[2] 'No closer relations of any kind can be traced between the oldest Angiosperms and the Gymnosperms. Both chief sections of flowering plants are as sharply separated in their fossil types as they are as living plants.'

(4) It should also be considered that as regards the history of the Mosses and the oldest Thallophytes we know practically nothing, despite that the conditions of preservation for the associated mosses were favourable, growing as they often did in damp and humid habitats.[3] Hence it is seen that regarding the historic (phyletic) development of the flora we can say nothing with certainty.

(5) The question whether within the limits of one and the same type—for instance the Lepidophyte class—the higher orders appear after the lower cannot be determined in the absence of objective evidence. It appears, however, to be very probable that the changes

[1] Reinke: *Naturwiss. Vorträge*, Vol. I, p. 28.

[2] Reinke: *Philosophie der Botanik*, Leipzig, 1905, p. 135.

[3] Gothan: *Entwicklung der Pflanzenwelt*, p. 96. Even in the Carboniferous formations mosses cannot be clearly made out. 'The question of fossil mosses, owing to these circumstances, has been a source of much brain-racking among the palæontologists.'

which the Lepidophytes, or the Equisetæ for instance, experienced in the course of geological periods represent a simple change of form, a specialization of one and the same grade of organization in varied directions. The reasons for this we will at once present.

§ 3. *Description of changes actually observed and the probable causes of same* (' *Palæontological Law of Evolution* ').

The transformation of the plant world is much less than is generally supposed. It is true that we read in all tuition books that the Coal flora, the New Red Sandstone flora, and others, are fundamentally different from each other and from that of the present day. In the wall cartoons of Potonié this is shown in the clearest way. These pictures show exactly the flora actually found in the formations concerned, but as regards the extent of the transformation established in single definite groups a mere comparison between two or more such ' landscapes ' can teach us very little.

This is because, first of all, in most cases it is only parts of the flora concerned which are shown—for instance, moor plants: consequently moor should be compared with moor. Furthermore, the climate at the time concerned must be considered: the tropical flora of the Coal era has its descendants mostly in the tropics. Finally it must not be forgotten that plant groups can die out and become extinct and, indeed, have done so frequently.[1] It would, then, be vain to look for representatives of such types. The presence of a single group of this kind, especially when it is numerically strong, naturally gives to a formation quite a different appearance in comparison with all others in which the group is lacking.

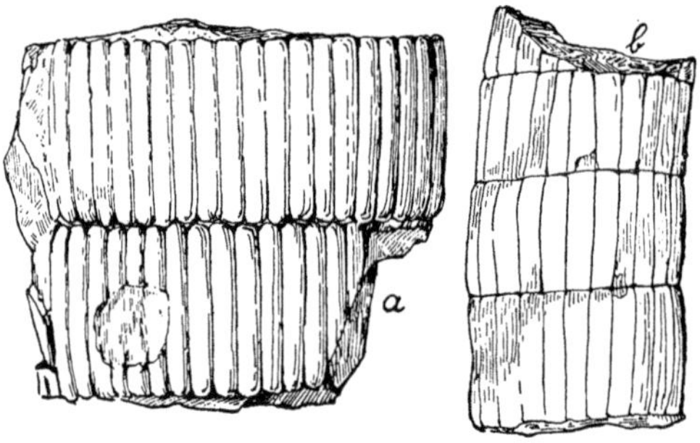

FIG. 19.—*a*, CALAMITES *Luckowic.* *b*, ASTEROCALAMITES *scrobiculatus.* Kulm. *Astero-calamites* belongs to the Protocalamariaceæ and shows the course of the furrows as described. (*After Gothan.*)

If this be borne in mind, then the transformation of the plant world loses much of its 'magnificence.'

With this preliminary remark we will, by several examples, show of what kind are the transformations observed and to what causes they may be imputed.

[1] As such extinct plant forms there are regarded for instance the Cordaites of the Coal Age, and mostly also the Lepidodendron and Sigillaria and other smaller sections. Steinmann (*Die Geolog. Grundlagen der Abstammungslehre*, 1908, p. 20) will not hear of 'dying out' in the sense of actual disappearance. Therein he goes too far, since the extinction of extremely specialized forms involves no improbability.

(1) *The Law of Specialization* (*Differentiation*).

(*a*) The Calamariaceæ Series.

In the Devonian formation, probably already in strata which may be ascribed to the Silurian period, there are found the first remains of Equisetæ—the so-called Protocalamariaceæ. In the upper productive coal formations these primary Equisetæ become Calamariaceæ. How did that occur? The Protocalamariaceæ show clearly, in the stone kernels preserved, the impressions of the main vein fascicles running upwards in the stem in the form of longitudinal furrows (Fig. 19) which, as opposed to those of the Calamariaceæ in the separate nodes of the stem, lie exactly in line with each other; in the Carboniferous forms, on the other hand, each furrow ends between two of the furrows of the upper node. Now it can be accepted as quite certain that this alteration had a definite purpose because, as Haberlandt in his excellent work[1] points out, the formation, position, and direction of the vascular bundles stand in the closest possible relation to the physiological needs. To what new life conditions these old Equisetæ conformed thereby we naturally cannot say with our present knowledge. This question can only be solved by constant consideration of the present 'adaptive evidence.'

If we now consider that the Protocalamariaceæ groups only occur in the oldest strata of the coal

[1] G. Haberlandt: *Physiologische Pflanzenanatomie*, Leipzig, 1909, p. 338.

formations and in the Devonian[1] formation, widely extended over the world it is true, but only in a few and similar forms, while the Calamariaceæ appear in several families and individually in greater number, we come near to the conclusion that under the luxurious conditions of existence in the Carboniferous era[2] a wide development of the Calamariaceæ type took place in various directions according to the nature of the habitat. That would be a case of differentiation of a type.

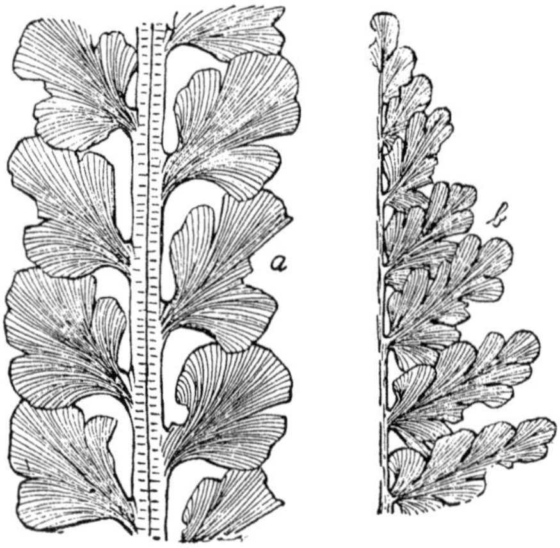

FIG. 20.—*a*, SPHENOPTERIDIUM *furcillatum*; Silurian; Heffen, Nassau. *b*, S. *dissectum*; Kulm; Rothwalthersdorf, Lower Silesia. Primary venation. (*After Gothan.*)

(*b*) The Fern Series.

The classification of the fossil ferns is effected for purely practical reasons according to the 'venation' and according to the mode of attachment of the pinnæ (subdivisions) to the stalk. The oldest ferns, the Archæopteridæ, show a fan-like venation; all the veins are of the same thickness and radiate from about one point[3] (Fig. 20).

[1] According to Zittel: *Handbuch der Paläontologie*, V, p. 176. Gothan puts a note of interrogation against Devonian.

[2] Potonié: *Die Entstehung der Steinkohle*, etc., p. 152.

[3] Gothan: *Entwicklung der Pflanzenwelt*, p. 21.

With the chronologically later ferns the venation becomes feathered, i.e. there are continuing main veins with lateral branching minor ones (Fig. 21).

In the rich Carboniferous flora there then appear fronds with reticulated venation in which the lateral branch veins are united together by numerous short

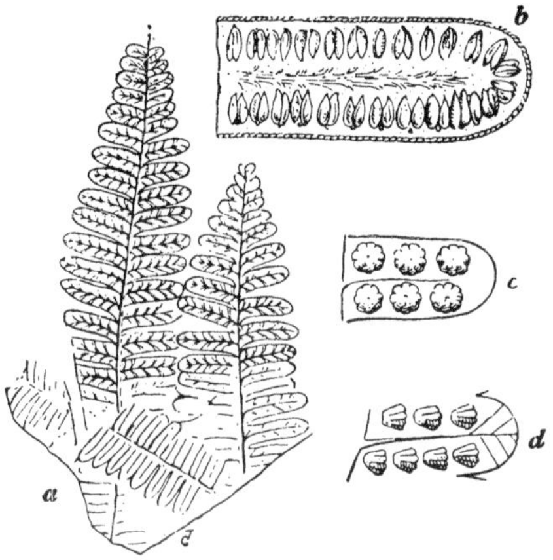

Fig. 21.—*a*, Pecopteris *oreopteridia*, Saar district. *b*, *c*, *d*, Fronds (parts of) of various ferns, showing spore heaps on the under side. (*After Gothan.*)

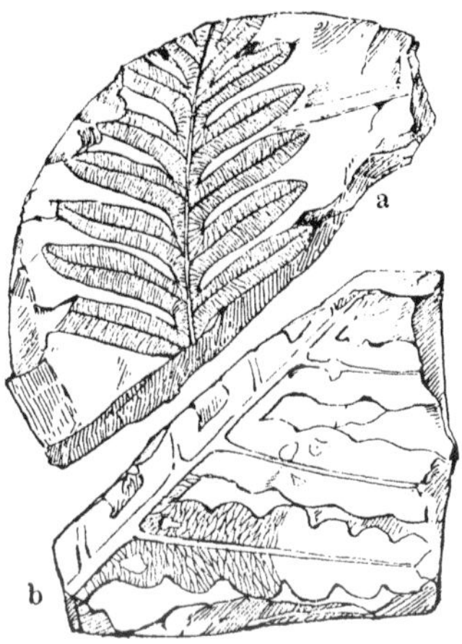

Fig. 22.—a, Alethopteris *Serli*, Saar district. b shows reticulate venation. (*After Gothan.*)

connective ones (anastomosing) (Fig. 22, b). From the purely comparative point of view the reticulate venation is decidedly an advance over the feathered venation, and this, in its turn, over the fan venation.[1]

[1] Haberlandt: *Physiologische Pflanzenanatomie*, Leipzig, 1909, p. 348. Haberlandt describes two chief types of venation in leaves. The first, in which the veins proceed separately (without anastomosing), appears as a rule in such leaves as never require much water or nutriment on account of their smallness or trifling transpiration, and show assimilative activity. The second type (reticulated) appears in leaves of the opposite character. 'Thus (by this venation) the leaf area, with the least possible length of veins, becomes uniformly and by the shortest way supplied with water and nutritive salts' (p. 349).

Since now chronologically also, from the Silurian to the upper productive coal measures, the said evolutionary forms follow each other, therefore the Ferns appear to present an evolutionary series and the preliminary of the transformation itself to have been a differentiation of the original Archæopterid type. The external cause we cannot discuss, since with the palæontological evidence we have only to compare the results of vital processes with each other.

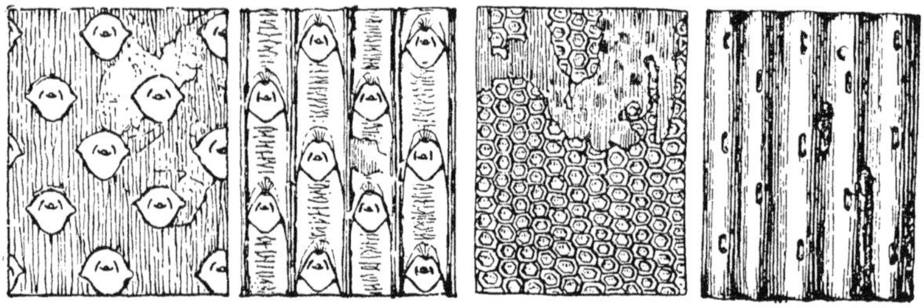

a *b* *c* *d*

FIG. 23.—*a*, SIGILLARIA *Brardi* (leiodermic) from the upper productive Carboniferous formation of Wettin. *b*, S. *Blobay* (rhytidolepic) from the middle productive Carboniferous formation of the Ruhr district. *c*, S. *elegantula* ('*favularic*,' ex same). *d*, SYRINGODENDRON (state of preservation). *c* and *d* show how different states of preservation occur by removal of separate bark layers. (*After Gothan.*)

(*c*) The Sigillaria Series.

A similar transformation, though caused by changed external conditions, we can follow up with the fossil Sigillaria.

According to the arrangements of the leaf scars the Sigillaria have been classified as 'favularic' when they stand quite close together in sloping lines, 'rhytidolepic' when they appear in longitudinal rows with furrows between, 'leiodermic' where the scars stand quite separated on the bark without longitudinal divisions (Fig. 23). In the said succession the Sigillaria form actual guides [1] for the various successive geological horizons. In this way they divide themselves and form as a whole a connected series by descent. The actual causes and the purpose of the described transformation must be deduced from observations of the plants at present living.

The Ginkgos show, particularly in the Jura formation, a great multiformity in the make of the leaves which are of value as indicative characters for the various systematic species and genera (specialization). In a general way separate links may be determined in the transformation of many other families which eventually have led up to the present ones, but regarding the details of the process we are not sufficiently instructed.[2]

(2) *Phenomena of Convergence* (p. 45).

By convergence is understood the formation of like or very similarly constituted organs, or, if the organs already exist, of similarities of structure and form in organisms which according to their total type belong to different systematic categories. Under some circumstances the limitations of such converging groups may naturally be rendered difficult.

It has long been remarked that, for instance, the Carboniferous flora appears outwardly of a fairly

[1] Gothan: *Entwicklung der Pflanzenwelt*, p. 41.

[2] Reinke: *Philosophie der Botanie*, p. 136. 'Actual transitional forms between Tertiary species and living species cannot be followed up with the desired clearness.' (Reinke speaks here of flowering plants.)

uniform stamp, which can easily induce the idea of close relationship and in some cases has even led to the putting forward of mixed types or transitional forms.

Now, however, we can impute many peculiarities of the great groups of the Carboniferous flora to convergence without conflicting with the facts observed. The increase of thickness of the stem by means of a constantly present embryonal tissue (Cambrian) in many of the Lepidophytes and the Calamariaceæ, the possession of large subterranean rhizomes [1] (Stigmaria),[1] the club-like thickening of the stem below, the possession of a smooth trunk without bark, the formation of stomata on the stem or in organs proper (Pneumatophora), can—nay, must—be regarded as adaptive factors to the like environmental conditions. The proof of this has been provided by Potonié in his work already mentioned [2] in the most convincing manner by the comparative study of the Carboniferous flora and that of an existing tropical marshland. Independently of the systematic differences between the fossil and the present moor flora—in the Carboniferous era Pteridophytes and Gymnosperms and, in the equivalent tropical moorland of Sumatra, Dycotyledons—the peculiarities of both flora are unequivocally explained by the mode of life under the same conditions.

[1] Rhizomes (root-stocks) are extensions or thickenings of the underground stem, from which the roots proper issue. The 'stigmaria' are very widely projecting twice-branched formations, whose surface is covered with round scars (whence the name stigmaria) spirally arranged, which stand far apart from each other, and sometimes bear round rootlets. E. Frans: *Der Petrefactensammler*, Stuttgart, 1910, p. 48.

[2] *Die Entstehung der Steinkohle, u. s. w.*, pp. 152, 166, and 169 (tables).

It does not therefore follow from the fact that in the Carboniferous formation many ferns formed actual trunks, that they are really related to the Gymnosperms (Pteridosperms). On the other hand there may be Gymnosperms (the said Pteridosperms) with fern-like foliage.[1]

In the coal strata of the Tertiary formation the flora of the present-day moors is much more similar, 'since these formations contain many genera and species of plants which still exist.' In North America we have to-day 'very extensive moors with a plant community of which a considerable number of species are the same as, in the Miocene period, occupied our moors.'[2]

(3) *Phenomena of Retrogression* (p. 44).

If our present Equisetæ be compared with the Calamariaceæ of the Carboniferous period, which are closely allied, a general degeneration of those gigantic tree-like growths might be assumed. In nearly all educational books indeed we find the same observation —that our Equisetums, as small insignificant weeds, of such great uniformity of make that all the species only form one genus,[3] are the degenerated dwarfed descendants of those fossil trees. If that be so, we

[1] In this case it is, in addition, quite immaterial whether the fern type of the leaf be regarded as evidence of convergence or not, since even to-day Gymnosperms exist with such foliage—the Cycads (Fern Palms)—which, however, are not therefore regarded as 'intermediate forms.' Oliver (*Biol. Zentralblatt*, 1905, p. 403) constantly speaks of 'intermediate forms' (*Zwischenformen*).

[2] Potonié : *Die Entstehung der Steinkohle*, etc., p. 185.

[3] Warnung : *Systematische Botanik*, p. 151.

have certainly here a good example of a general retrogression before us. That would not be an impossibility, since, supported by the comparative study of the present moorland flora and the fossil one, the thickening of the stems, the abundant ramification of the lateral growths, the form of the leaves, which are not increased in area but only split up (Fig. 24), may all be imputed to the influence of the favourable environment in the Carboniferous period. If such conditions cease, naturally all the special conforming structural arrangements cease also. Therefore the lack of thickened stems, the reduction of leaf area, and the reduced branching of the present forms, can scarcely be called an actual degeneration. But in the Carboniferous period there were in all probability weed-like small forms as well as the gigantic ones;[1] and from these we might well attempt to deduce the modern Equisetæ. Certain it is that the existence of small club-mosses is proved.

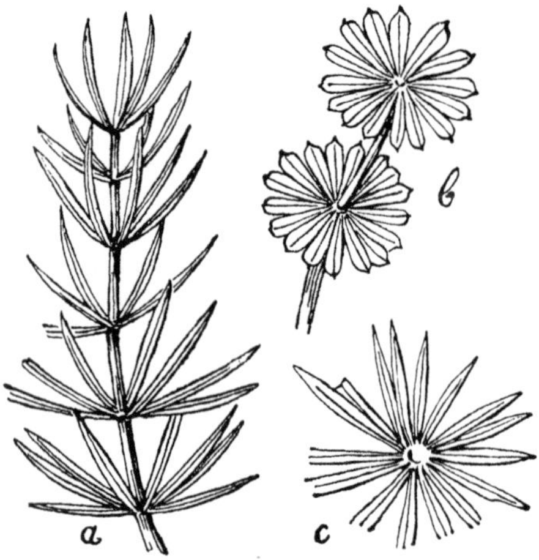

FIG. 24.—CALAMARIA FOLIAGE. (*After Gothan.*) *a*, *Asterophyllites equisetiformis*; Carboniferous, Harz. *b*, *Annularia sphenophylloïdes*; Zwichau. *c*, *A. radiata*; Lower Silesia. The leaves have not assumed scale-like forms as have the present ones.

[1] Gothan: *Entwicklung der Pflanzenwelt*, p. 51.

The Calamites we no longer find in the following formations, but throughout the whole Mesozoic formations we find true Equisetæ (like the modern ones with slender branched foliage), as also in the Permian, Red Sandstone, Jura, etc., which alter but little. Among these are large forms, of which we retain an example in the present *Equisetum giganteum*, ten to thirty feet high.

It is to be expected that from the outset retrogressive phenomena should play a considerable rôle in the history of the plant world and largely contribute to shape new forms, but it might be difficult to produce the necessary fossil proofs. We deduce this rather from the so-called 'rudimentary' structures of the present plants, particularly parasitic ones.

Conclusions deduced from palæolithic botany.

If it be granted that the forms of growth of systematically higher rank appeared chronologically subsequently to the lower, yet there is absolutely no indicative proof in the case of any one group (family or class) that they were developed from the lower forms. In this botanists are entirely in agreement in so far that no one speaks of an actual 'proof' which can be produced in the shape of fossil evidence of transitionalforms.[1]

[1] See, for instance, Zittel's great work, the fifth volume of which is elaborated by two distinguished botanists, Schimper and Schenk. The expressions used of 'possibility,' 'perhaps,' etc., evidently show that no proof has been afforded. Furthermore, see Reinke, Schwendener, Haberlandt, Potonié, Gothan, Steinmann, Kothen, Depéret, Kerner v. Marilaun, and Neumayr. Certainly in many works we must differentiate between what the authors represent as actual results of investigation in their special lines and what they add thereto regarding 'general problems.'

On the other hand it is probable—in some cases very probable—that by adaptation to varied environments a type which has once appeared branches off into divisions of varied appearance—as with the Lepidophyton of the Carboniferous period for example. We must therefore accustom ourselves, with plants as with the animals, to speak rather of a 'transformation' and alteration of form than of an actual higher evolution.[1]

[1] 'Higher evolution,' in the strictest sense of the words, would be correct if (1) an objectively based division into higher (more perfect) and lower (more imperfect) grades of organization of the animal and plant kingdoms be accepted, and (2) if such a higher grade has been formed from such a lower one. The second premiss we must dispute, the first we will discuss hereafter.

SECTION II.

THE EXPLANATORY DOMAIN OF THE HYPOTHESIS OF EVOLUTION.

CHAPTER I.

INTRODUCTION.

According to what we have so far stated in connection with palæontology the object of an evolutionary or transformation hypothesis is fairly well defined.

In the first place we have to inquire, by observation of, and experiment with, the organisms of to-day, whether they are generally capable of transformation, what causes are thereby involved, and of what kind are the changes ascertained.

Thereby we arrive at the first and entirely indispensable basis of any attempt whatever at scientific explanation through observation. The second part of the task involved would be to imagine the same causes as effective in the past, alone or in connection with other influences of similar kind, and then to compare the chronologically successive organisms of ascertainable form and structural conditions with those still subject to observation. If both show the same peculiarities, then we may conclude with perfect right that the modifications of the primeval animals and plants were really brought about by those causes or, better expressed, were induced by them. If the changes of form of the fossils remain within the limits of those alterations which we at present observe in the recent organisms, or can, with great probability, deduce from them, then are we certain that our explanation is correct. If they extend farther, we must inquire whether an increase in the intensity and the duration of that influence may not explain the great scope of the deviations. The certainty of our deductions certainly is decreased thereby. How far the application of these principles may be carried is not, however, left open to choice which might be satisfied with a mere glimmer of possibility and probability: we must act within the limits set by Nature and by Science, regarding which we have already said what is needful when discussing the results of palæontological research.

This truly scientific standpoint is that assumed by many eminent palæontologists, such as Neumayr, Waagen, Zittel, Koken, Steinmann, Depéret, Kerner, Marilaun, Reinke, and Wasmann.

We believe, however, that in many cases, especially in advanced 'reading circles,' such a standpoint is regarded as simply 'naïve': one is accustomed there to see quite other and deeper-seated questions treated after a certain 'dogmatic method,' in books on evolutional history, which commence with these purely scientific problems and proceed to the most subtle questions of world-wide breadth (*Weltanschauung*). Why we do not do that, we will explain as follows.

In the first place, a word on the so-called 'palæontological methods.' This is certainly that method which adheres to the historical development of the entire evolutional problem, and therefore also that which repeats the original thoughts and permits the newly introduced extensions thereof to be determined. This was the method adopted by an eminent palæontologist —Ch. Depéret—in his work so frequently mentioned, and the honourable reception which the book also experienced among German savants[1] shows that no one found any objection to the plan adopted.

This, however, does not imply that palæontology is alone called upon to finally determine the limits of transformation. Thus the possibility of the descent of one plant type from another, despite the perfectly negative results of palæontology, is by no means disproved. In many cases a decision cannot be arrived at by the study of fossils alone because, for instance, the petrifactions of the pre-Cambrian formation—even if they existed in larger numbers—are in any case shattered or destroyed, and indubitable plant remains are not so far known from pre-Silurian formations. Under these circumstances what can be said regarding the earlier history of the Ferns or the Trilobites? It will, however, well be conceded that it is inadmissible to speak of and appreciate only the favourable side of the results of palæontological research—viz. the abundant evidence in favour of manifold new forms—and to be silent regarding the lack of any probable tangible

[1] The book was very favourably spoken of by Frech, Koken, Steinmann, R. Hoernes and other palæontologists.

evidence of the transition of one family into another, or one class into another, both in the animal and plant kingdoms. Both must be considered in the same way, unless the other indices for the evolution of organisms—comparative anatomy, embryology, animal and plant geography, etc.—lead to such strong indirect proof that the negative results of palæontology in separate cases, or even generally, may be ignored. In this case the theory or hypothesis could really serve as an explanation of an occurrence without the occurrence concerned being a support to the hypothesis.[1]

Then, for example, the conclusion that, despite everything, a definite class must have been derived from another, could, considered purely theoretically, become an actual scientific postulate.

The 'scientific postulates'—here the theory of evolution—we meet with now very often in contemporary literature. Many of these we cannot at all recognize as such, since a condition absolutely essential for the statement of a postulate is left out of consideration. For instance, no trouble is taken at the outset to ascer-

[1] Only in this sense is Naegeli's expression correct : 'It was not my intention to discuss all branches of the doctrine of descent . . . Therefore the otherwise generally treated theme of geographical extension and the palæontological evidence have been almost entirely neglected by me, because the existing demonstrated facts show themselves to be of manifold significance, and because their explanation may be much rather expected by means of a correct theory than that they should contribute appreciably to the foundation of one.' (*Mechanisch-physiolog. Theorie der Abstammungslehre*, Munich and Leipzig, 1884, v.). Naegeli in point of fact proceeded in the said work on purely *a priori* lines, since what 'is' he spoke of as 'becomes,' expressing actual facts connected with the present organisms in evolutional historical fashion.

tain whether the facts concerned really come within that category, from which alone the basis of an evolutionary hypothesis can be formed. An example in another direction will explain what we mean. Let it for instance be conceded that the physicists justifiably claim the existence of ether and that of definite vibrations of its smallest molecules because thereby they obtain a satisfactory conception of the phenomena of light, electricity, etc.; therefore it is scientifically perfectly justifiable to employ the hypothesis of the ether experimentally also to account for other natural processes, provided of course that the new batch of facts requiring explanation can be comparable with the phenomena of light and electricity.

But no one could seriously attempt to use the ether hypothesis in order, say, to explain consciousness, memory, and the will, since consciousness by its entire nature has absolutely nothing to do with a material state of vibration. The basis of such an attempt could be formulated in the following way: 'Since, by the recognition of a material ether and of definite vibrations of its molecules, we have received an acceptable idea of light, the rapidity of its transmission, refraction, etc., therefore it is a scientific postulate that there be recognized also consciousness, and action—in short the said capacities—as ethereal vibrations, even though all possibility is lacking of saying how it is done.'

Unfortunately the formulating of many evolutional 'postulates' is similarly framed!

It is a quite inadmissible procedure to put forward

things and processes in the explanatory formula of an evolutionary hypothesis as the postulate of the same, when such things or processes cannot be recognized in their entire nature as the results of an evolutionary process. That would be the case if, for example, in the whole of our experience of the accessible world absolutely nothing could be found which reasonably could be regarded as a foundation, as an undeveloped 'latent' form—in short, as a beginning of that which it is desired to explain, since 'evolution,' in itself, signifies development and extension of a thing or condition which at least can be suggested as existent. If two things—of which the one has the perfection or capacity under consideration, the other possessing neither even as a commencement—exist together, these may have manifold relations to each other and naturally may affect each other, but through 'evolution' they have no connection.

CHAPTER II.

LIMITATION OF THE EXPLANATORY DOMAIN OF THE EVOLUTIONARY HYPOTHESIS.

Preliminary Observation.

THE opinion is generally held that the natural objects which surround us may, in the first place, be divided into two great and quite different groups—the animate and inanimate. The animate, again, are divided into the so-called animals and the plants, and these are treated as separate branches of natural science. The animal kingdom and that of the plants are again divided by systems, agreeing in their main features, into stocks, classes, orders, families, genera, species, and sub-species, etc.

It is therefore indubitable that, for the acceptance of such a threefold division in the things themselves, some sort of starting-point must exist, otherwise it would have been impossible to establish associated but strictly separated scientific branches, since the definition of those branches is effected according to the difference of their objects.[1]

[1] It can well be said that most students of nature, in such division of the natural objects into separated groups, perceive the expression of actual relationship. Otherwise there could not be understood the standpoint of the chemist and physician with regard to 'pure' biological questions, and that of the physiologist and biologist with regard to 'pure' chemico-physical ones. The physician considers, because he leaves the construction and activity of organisms to the biologist and physiologist, that he is thereby limited in *his* particular domain, and *vice versa* the students of the organic branches have their own methods and domain of investigation.

According to many evolutional theorists these contrasts between the organic and the inorganic (animate and inanimate), between animals and plants, between families and classes within the same kingdom, should not be of such a kind as to be inexplicable by 'evolution.' The acceptance of a genetic connection is indeed a 'postulate of the evolutionary doctrine.' This we must contest, since there are lacking—at least so far as the origin of life from the inorganic world and the evolution of animals from plants are concerned—all the conditions for the acceptance of a 'postulate.'

§ 1. *We are not justified in regarding the origin of organisms upon our Earth as the result of an evolutionary process.*

It is no part of our task to consider here all the attempts which have ever been made to explain the appearance of life upon our planet. Fechner and W. Preyer, for instance, accept the priority of life and deduce the inorganic from the organic. According to them the lifeless bodies were 'the signs of the dead primeval gigantic organisms, whose breath perhaps was glowing iron vapour, their blood molten metal, and their food perhaps the meteorites.'

Several eminent physicists—Helmholz, W. Thomson (Lord Kelvin), and most recently Svante Arrhenius—represent the 'opinion' that organisms have ever been associated with the inorganic material, and, pervading the universe, spread the germs of life wherever a world anywhere became capable of providing a habitat for organic existence.[1]

This idea owes its origin to the 'impossibility that lifeless material can pass into living.'[2] The investigators named do not therefore rank with our opponents: their opinion implies rather 'a fundamental difference between living and inorganic substance and a duality *in infinitum.*'[3]

Such attempts at explanation have naturally not met with much approbation because they belong to the 'merry realm of speculation' and are absolutely beyond proof.

Now and again the belief arises in the doctrine of 'spontaneous generation,' the most acceptable attempt at explanation of the origin of life on our planet. This implies the spontaneous generation of living bodies (organisms) from the ordinary (inorganic) materials as the effect of ordinary chemical and physical powers, either under special or also under the conditions ruling at the present time.[4] As a rule 'special' environmental conditions are demanded which do not now present themselves, but no details of such specializations are given.

[1] Reinke: *Die Welt als Tat*, Berlin, 1905, p. 344. See also E. v. Hartmann: *Das Problem des Lebens*, Bad Sachsa i/Harz, 1906, p. 178; and H. Muckermann: *Grundriss der Biologie*, I, Freiburg, 1909, p. 144.

[2] O. Hertwig: *Allg. Biologie*, p. 272.

[3] Reinke: *Die Welt als Tat*, Berlin, 1905, p. 345.

[4] According to Naegeli (*Mechan.-Physiol. Theorie der Abstammungslehre*, 87) this generation occurs in the 'warmer seasons' even to-day and in our own regions, principally, however, 'in the warmer climates, and in the old primeval time after the cooling of the earth down to breeding heat.' Usually the possibility of spontaneous generation is only attributed to the earliest times.

Such an origin of organic life must be denied in the name of Science, because

(1) *Between organisms and inorganic material there is an essential difference, so that the inorganic material cannot develop itself into an organism.*

For the comprehension of our argument the meaning of the words 'organism' and 'life' must be examined more closely. The simplest and most general definition which is afforded by modern biologists is as follows:

> An organism is essentially a whole composed of material and functionally varied parts.

Both elements of the signification—manifoldness and unity—are clearly expressed in the word 'organism.' The ending 'ism' points to a collection of organs. 'Organ' is a tool for a defined service: all tools together form the one whole.

The peculiar way in which this whole is in itself active is termed 'life.' It can be briefly defined as 'the definite co-operation of all the limbs (parts) determined by constant consideration for the whole.' So comprehended this definition suits all natural bodies which are usually considered as living,[1] even the simplest

[1] There are other and very good definitions of 'Life.' Scholastic philosophy particularly has very thoroughly treated the doctrine of life. Since, however, we have desired to touch upon the problem of life only in so far as is necessary for a critical examination of its origin—as it is presented by many modern naturalists—we may content ourselves with brief indications. For the use of many very indefinite expressions we must hold the naturalists concerned responsible.

which we so far know—the cells—whether they be individual and independently existing beings (monocellular), or constituents of a so-called ' higher ' organism. Since for the first time Brücke (1861) has demonstrated the ultimate morphological and physiological units of which the organs in animals and plants consist—viz. the ' cells ' —to be an aggregation of varied kinds of parts connected together by definite laws, the organic cells have also been designated as organisms—elementary organisms.

This term is very unhappily chosen, since a liver cell or a kidney cell cannot be termed an ' organism ' in the sense of an independently existing and active whole.[1] Since, however, the expression has become widely spread, and used especially by most of the defenders of spontaneous generation (*Urzeugung*), we will accept it for the time being. It is important for us to remember that the cell—that is, the absolute or relative ultimate unit of living matter itself—is composed of many varied elemental parts which, like the organs in a higher organism, co-operate in the vital process.[2] If, therefore—precisely because the cell consists of many varied and co-operating parts—it has been termed an organism, that then is a proof that our definition is accepted by the biology of to-day. We can, therefore, proceed upon that basis.

In the definition of ' life ' we spoke of the ' definite co-operation of all the parts determined by constant

[1] See E. Wasmann: *Die moderne Biologie und die Entwicklungstheorie*, p. 190.

[2] O. Hertwig: *Allgemeine Biologie*, II.

consideration for the whole.' That is shown by the fact that each part works for all others precisely as it does for itself: thus the lungs breathe for all the tissues, and digestion is effected for the benefit of all, and so on.

Thereby each organ is itself also dependent upon the normal execution of all other functions. If there arise an increase of activity in any organ that is so because the whole or some particular part requires it. In short there is shown clearly in every part a striving to remain preserved in conjunction with the whole or the necessity of perishing with it.

M. Heidenheim[1] very beautifully describes this relation of the separate components to the whole: 'In the more highly organized creatures the cells present themselves as subordinate parts of the whole, which have lost the freedom of existence and of action, a condition which is designated by H. Spreuzer as an "integration" of the individual, because it has become an integral part of the whole. In connection therewith there is the extremely varied differentiation of the specifically functional cells. . . . It is this change of constitution which, in the theoretical sense, confirms and determines the dependent relations of the whole, the integration of the individual and the subordination of the cells to the position of mere tools which serve for the vital work—the "life."'[2]

[1] *Plasma und Zelle*, 8th volume of *Handbuch der Anatomie des Menschen*, published by Bardeleben, Part I, Jena, 1907, p. 29.

[2] Heidenheim contests in this work with powerful arguments the doctrine that a complex organism is a cell community (*Zellestaat*): see particularly p. 49.

From the above it is understood, when it is said of the organisms, that they constitute a purpose in themselves. Observation indeed shows us that they attain no other object than self-preservation or increase in number by the reproduction of like beings. We see the same if we more closely observe their behaviour towards inorganic material and the universally effective powers of Nature. The organisms, it is true, are thoroughly dependent for their existence on the chemico-physical powers and the inorganic materials, but their relation to these is so regulated that they utilize and profit by such influences and effects as are consistent with their own maintenance.[1] Furthermore the living bodies cannot effect any material work at all independently of the material and the peculiar powers which accrue to it also from outside; but the chemical materials, for example, are so arranged, and in their activity so ruled, that they only act in such a manner and at such a time, how and when, as is necessary to the organism for its purposes.

From this generally conceded dependence upon the use of material energies it might be concluded that the organism, after utilizing same, must come to a standstill; but that is not so. The living body obstinately maintains itself in its active condition, since it rejects the exhausted material and seeks and assimilates a new supply. With the fresh material it forms highly

[1] Naturally, however, we will not deny that all organisms can be destroyed; they are not *absolute* existences. The purposeful 'utilization' of the inorganic materials and powers is evident in the normal life—that suffices perfectly.

complex and highly changeable chemical combinations and thus creates new springs of energy. This continuous exchange of material, by which nothing else is attained and striven for than the self-maintenance and multiplication of its individuality, shows itself in the most striking way in the external appearance and is therefore utilized by many investigators in their definitions. Thus, for instance, O. Hertwig states: 'Life (using a general expression) displays itself therein that the cell, by virtue of its own organization and under external influences, suffers constant changes and develops powers by which its organic substance, on the one hand, under definite expressions of energy is destroyed, and, on the other, is again renewed.' [1]

All that has so far been stated shows that the organisms possess a power of their own to strive to attain certain purposes and that they only convert and use inorganic material in their own interest.

This conception is entertained also in one or another form by the majority of modern biologists,[2] as, for instance, by Cl. Bernard,[3] O. Hertwig,[4] K. E. v. Baer,

[1] *Allgemeine Biologie*, p. 65.

[2] Christian philosophy has always taught this and has never accepted that enigmatical 'vital force' which should be a particular form of energy, but has conceded to the organism a substantial principle as having a purposeful striving power.

[3] *Leçons sur les Phénomènes de la Vie*; I, Paris, 1879, p. 51. 'The vital force (here = cause of unit direction) directs phenomena which it does not produce; the physical agents produce phenomena which they do not direct.' II, p. 524: 'The ultimate element of the phenomenon is physical, the arrangement is vital.'

[4] *Allgemeine Biologie*, pp. 16, 18, 65. The same: *Der Kampf um Kernfragen der Entwicklungs- und Vererbungslehre*, Jena, 1909, pp. 75, 80.

Heidenheim, Pfeffer,[1] Bunge,[2] Reinke,[3] H. Driesch,[4] Strasburger,[5] and practically and actually by all botanists and zoologists, even those who theoretically still regard life as a chemico-physical process of peculiar complexity.

Conclusions from the above.

1. 'Life,' in the next place, is understood quite generally by modern naturalists as a peculiar mode of activity which we only find in the so-called organisms, i.e. in systems which exteriorly are contained within themselves and which consist of chemically varied and structurally differentiated parts. It consists essentially therein that all parts of such an organism naturally act together or are active as instruments in the service of the whole and on their own initiative.

That there are natural bodies which display the activity described, in contrast to others in which it does not exist,[6] is, for Cl. Bernard, 'a fact to be no longer disputed'; for O. Hertwig, 'a fact and

[1] Pfeffer's decision in the *Pflanzenphysiologie*, Leipzig, 1897 and 1904, we will give hereafter under the discussion on stimuli.

[2] *Lehrbuch der Physiologie des Menschen*, II, Leipzig, 1905, p. 7. 'In the activity, therein lies the enigma of life.'

[3] *Einleitung in die Theoretische Biologie*, Berlin, 1901, sections 3 and 4; also *Die Welt als Tat*, Berlin, 1905, chaps. xxii. and xxiii.

[4] We will treat later on of Driesch in detail.

[5] The well-known *Lehrbuch der Botanik für Hochschulen* maintains this opinion and establishes it excellently.

[6] Cl. Bernard (*Leçons sur les Phénomènes de la Vie*, p. 50) calls the 'wonderful subordination and the harmonious co-operation of the vital activities' a fact to be no longer disputed ('le mot importe peu, il suffit que la réalité du fait ne soit pas discutable'). O. Hertwig (*Kampf um Kernfragen*, etc., p. 80): 'That the living substance . . . in accordance therewith (i.e. with its organization), develops peculiar ways of working, is, in my eyes, a fact and no mystical conception as Verworn regards it.'

no mystical conception as Verworn regards it.'[1] In an organism all parts thus show 'final change relations' (*finale Wechselbeziehungen*) which are lacking in inorganic material.[2]

2. The result which is attained in all cases by this co-operation is exclusively the maintenance and reproduction of the organism itself. This result is rendered possible by the capacity of the organisms to utilize and assimilate the inorganic material whenever it is required. This capacity is shown, for instance, in the purposeful selection of the material which normally is taken up, in the exclusion of poisonous matter[3] and the ejection

[1] Verworn maintains—it is true, casually—in *Allgemeine Physiologie*, Jena, 1901, the essential similarity of the vital processes and the chemico-physical ones, but he thinks in the next place only of the 'bodily phenomena of life' (p. 7); the physical form a problem of their own. On p. 106 he concedes 'that the living substance cannot be associated with the chemical without being killed.' Then is the life departed! The most superficial conception, at least in some places, we have found in the otherwise excellent work *Traité d'Histologie* (published by H. Prenant, Bouin, and Maillard), I, 4. Prenant must have had here an opponent in view, who has maintained all sorts of nonsense. On p. 18 he himself describes the peculiarity of life as the struggle for the maintenance of a type associated with constant protoplasmic change.

[2] C. v. Hartmann: *Das Problem des Lebens*, p. 206. In cell-like forms of inorganic nature 'each part is as it is, and must be, according to effective molecular local forces, but it is not a serviceable member of a higher whole. Between the parts there occur certainly causal, physico-chemical changes, but no final change relations by which each part serves all the others and all of them together minister to the whole.'

[3] G. v. Bunge: *Lehrbuch der Physiologie d. Menschen*, II, p. 5. 'We know that the epithelial cells of the bowel never permit the entrance of a whole series of poisons although these in the fluids of the stomach and bowel are quite easily dissolved. [Thus the mechanico-chemical conditions for absorption are determined!—Remark by author.] We know even, that if we inject these poisons direct into the blood they become, on the other hand, ejected through the walls of the bowel.' For other very strong proofs of the utilization of the natural forces, see the same, p. 3.

of the matter exhausted by the organism itself, in the increase and diminution of the functions proper to the whole or to the parts, with the consequent morphological changes of form—the so-called 'adaptations.'

A result or an object that under certain circumstances may be enforced through adaptations, is obviously one striven for, one with a purpose: the preservation of the organism is thus the peculiar purpose of life; or in other words, the organism is at once the bearer and the purpose of life.[1] It is the bearer because the possibility of life depends upon its organization; and it is the purpose because nothing else is attained and striven for than its preservation—of its individual self for a period, and of its species, so far as in it lies, for ever.

That, likewise, must be conceded by biologists and physiologists, since it is never a question of another result or purpose when the subject treated of in the textbooks is the vital properties of organisms[2] (or the cell, which is regarded as the simplest form of organism).

3. It is granted by investigators that all material products of labour which are observed in an organism result from utilization of purely inorganic energies. It is, however, equally certain that all material activity is directed to a single end—the preservation of the whole.

[1] Here it is a question of the so-called 'inner' (i.e. inherent) purpose in things which is striven for by them directly (*finis internus*). The question, Why organisms exist? would be one concerning their external purpose (*finis externus*).

[2] See, for instance, B. O. Hertwig: *Allgemeine Biologie*, p. 65.

Cl. Bernard describes in a classical fashion this double side of vital activity : 'La force vitale dirige des phénomènes, qu'elle ne produit pas, les agents physiques produisent des phénomènes qu'ils ne dirigent pas.' [1]

4. By these words there is a new deduction expressed, viz. that the conduct and constant regulation ('direction') of the inorganic powers must have a special course of their own. Cl. Bernard uses for this the words 'force vital' (vital force), but does not think of embracing therein any of the other known forms of energy of equal power, but can only identify them with the organizations themselves.

J. Reinke calls the cause (or causes) of the purposeful direction of the purely material energies 'Dominants,' and explains the term as follows : 'In the organisms work is done by energy—by the Dominants the work to be done by energy is determined.' [2] 'Their existence is therefore a necessity, because without them only purposeless forces, first hand (*erster Hand*), would be active ; we see, however, in point of fact, that both the chemical and the constructive processes in plants and animals proceed purposefully and in unison.' [3]

H. Driesch demands the acceptance of 'Entelechia.' The word 'Entelechia' signifies the inner conformity of living bodies ; in a wider sense, 'the actual, elementary

[1] Cl. Bernard : *Leçons sur les Phénomènes de la Vie*, I, p. 51. Vital force directs phenomena which it does not produce ; the physical agents produce phenomena which they do not direct.

[2] *Die Welt als Tat*, p. 292.

[3] *Ibid.*

natural agency which expresses itself in them.' 'Entelechia utilizes the factors of the inorganic in order to produce that which is suitable to the particular species concerned and to regulate its preservation.' [1] 'Entelechia is that something which carries its purpose in itself (ὅ ἔχει ἐν ἑαυτῷ τὸ τέλος).' [2]

Since we are not idealists, the 'Dominants' and 'Entelechia' do not signify for us any symbols and abstractions (Reinke), but actual things which we, with the philosophy of past ages, call the 'soul.' As evidence there suffices, however, the confession that in the activity of the organism there asserts itself a principle which stands above inorganic matter and forces.

An organism is thus a natural body which by virtue of a directive and regulating principle conducts material activities and products of such on a plan and regulates these actively and purposefully to an end which is inherent in the organism itself.

5. Since the inorganic material, left to itself, never betrays the trace of a tendency to form such a system in which the separate parts are only instruments, there is, between organisms and combinations of inorganic

[1] *Der Vitalismus als Geschichte und als Lehre,* Leipzig, 1905, pp. 242, 246. He treats this subject very fully in the *Philosophie des Organischen,* II, p. 137. H. Driesch appears to have made the study of life his particular task. His evidences of the autonomy of life in his many works and writings are executed with a marvellous expenditure of thought, but are inaccessible to ordinary men through the many new terms and his mathematico-analytical methods.

[2] *Philosophie des Organischen,* I, p. 145. The word is borrowed from the writings of Aristotle.

matter and their mode of action, a fundamental elementary difference.

A fundamental difference cannot be bridged over by evolution, which by its meaning indicates a need of a starting-point as basis for a perfection to be developed. Therefore spontaneous generation in the usual sense of the term is excluded. The inorganic materials can, taking them absolutely, perhaps form those chemical combinations which appear in the organism; but they cannot, by themselves (*sponte*), adopt a direction and a higher purpose, nor produce Dominants and Entelechia, because these stand above them and have nothing to do with material energy.

We have thus, in the activity of living bodies and in the behaviour of non-organized matter left to itself, learnt to know two kinds of natural phenomena, both of which are alike elementary (primary). Elementary natural processes cannot be deduced from each other: there can be no question of bringing living organisms and unvivified matter into genetic connection by evolution. Everything that we know of the origin of the present-day organisms agrees therefore entirely with this, viz. that the phrase *omne vivum ex vivo* (and *omnis cellula ex cellula*) stands unshaken—nay, is more firmly established than ever.

(2) *The attempts to demonstrate as possible a genetic connection between vivified and non-vivified matter, or even to describe the course of the process, must be regarded as perfectly vain.*

It is, in short, clear that no investigator considers that through chemical structures of the most complicated sort there can be explained the origin of a new substance or of a directing 'something' that stands above the properties of matter. Although this 'something' may be clearly equally effective as the purely energetical processes in the organism, and therefore formally demands a scientific explanation, it is only too often overlooked. The organism itself is alone regarded, and it is believed that thereby the second element of vital phenomena, the constant purposeful direction of the purely chemico-physical processes, has been traced to a sufficient cause. But that is not the case.

(i) *No organization, which is regarded only as a peculiar chemico-physical quality or structure of inorganic matter, explains life.*

The term 'organization,' as it is particularly used by O. Hertwig and Cl. Bernard, is, in point of fact, a purely biological one, and means 'the albumen bodies which build up the protoplasm and all its innumerable derivatives . . . and stand in nearer and ordered relations to each other and constitute the being of the organism.'[1] 'Organization' means, thus, 'ordered

[1] O. Hertwig: *Allgemeine Biologie*, p. 16.

relations.' It signifies also, even, according to Hertwig—and the same may be said of other naturalists who come here into consideration[1]—that it is not merely by the existence of a definite physical aggregate condition, or by a stiff internally connected mechanism, as is the case in a machine, or by peculiar chemical combinations, that life can be explained. That the phenomena of life can be referred to the properties of the liquid aggregate condition O. Hertwig denies 'emphatically' and refers in that connection to many other investigators.[2] The 'structures,' however, also explain nothing, since all the coarser structures observed in cells—tissues, threads, network structures, and seed structures—have shown themselves long since to be temporary forms, conditions of the protoplasm, or the morphological expression of a particular function. They can

[1] Even Prenant, the author of the first section of vol. i. of the *Traité d'Histologie*, can only classify the combinations of matter which take place in protoplasm according to their purpose and their vital importance. He defines six groups—foodstuff, reserve and excreted products, and rigorously specialized instruments, e.g. myelin, chlorophyll, hæmoglobin—supporting material and active material. Not a word of chemico-physical points of view which, for instance, arise in connection with aggregate conditions or definite molecular groups. He introduces this grouping by the words: 'On peut cependant [since 'chemical' is not concerned], établir une classification biologique parmi les différentes substances trouvées dans les cellules' (see p. 10). See also particularly E. B. Wilson: *The Cell in Development and Inheritance*, New York, 1900, p. 316. This work is certainly the best that has been written on general biology. Furthermore, O. Hertwig: *Allg. Biologie*, p. 26; E. Wasmann: *Moderne Biologie*, chap. iii.

[2] *Allgemeine Biologie*, p. 16: 'With Naegeli and many other investigators we share the conviction that the complicated phenomena of the vital processes—before all, those of inheritance—are not explicable by the qualities of liquids or matter in solution. Wiesner is justified, therefore, in terming the attempt to refer the peculiarities of the living substance to qualities of liquids a surprising one.'

be permanently together in different cells of the same organism and stand obviously in the closest connection with the separative function of the separate organs; they may arise successively in one and the same cell, according to varied functional conditions—for instance, times of rest and activity. This applies also to the structure of the nuclear substance.[1]

According to the latest views of the study of the cell we cannot speak of ' organs ' of the cell in the sense of formations permanently existing and indispensable for definite objects (e.g. cell division). That is true for the cell skin, the centrioli, the ' chemical central-bodies,' the nucleoli, and the rest of the included constituents, and even for the nucleus. It is no longer correct to refer to the nucleus, in the definition of the cell, as one of the equivalent parts of the rest of the cell contents. The nucleus does not appear at all in an enclosed bladder in many unicellular organisms, but only the so-called nuclear substance; during each cell division it is perfectly dissociated individually, but it is not essential that it has a definite structure. The cell is therefore that more or less exclusive and independent mass of vivified matter in which alone the vital functions are exercised. The cell assimilates, the cell divides itself, etc., the nucleus and the centrioli being only integral constituents of the total mass. Wilson expresses this as follows: 'A minute analysis of

[1] St. Maziarski: *Sur les changements morphologiques de la structure nucléaire dans les cellules glandulaires*, in *Archiv für Zellforschung*, IV, Leipzig, 1910, p. 443.

the various parts of the cell leads to the conclusion that all cell organs, whether temporary or permanent, are local differentiations of a common structural basis.'[1]

Furthermore, the acceptance of submicroscopic structures, for instance, in the sense of a connected filamentary structure as a mechanical basis of the vital processes, is of no assistance, since there cannot obviously be attributed to such hypothetical structures any such qualities, as for instance, of rigid 'mechanism' structure, from which the actually observed new formation and transformation of the so-called cell organs (centrioli, nucleus, filamentary structures in the protoplasm) could not result. The cell clearly disposes quite freely of its material, it builds from its common basis the centrioli, always completes its nuclear substance, forms new nucleoli and dissolves all this again according to whether it requires one of these organs precisely for a particular office or not.

Conclusion.—There is in the cell no rigid mechanical structure either in the protoplasm (cytoplasm) or in the nucleus. The 'organization' is in fact only the purposeful co-operation of all constituents of the cell contents (regarded as elementary organism) or the displayed subordination of all parts in the service of the whole shown in the activity of the cell; 'organization' is thus in the meantime a purely biological term, i.e. it expresses only a peculiarity of the activity of

[1] E. B. Wilson: *The Cell*, p. 327.

the organism, not a chemical formula or a material, rigid, mechanical system.

No conceivable organization is by itself a sufficient cause of the purposeful direction of all the constituents of an organism and of its power, at first hand, for the best benefit of the whole. So far as a structure is observed in any way, it shows itself as the effect of a function. The functions are the primary factors, or rather the whole which utilizes material matter and energies for the common benefit.

What, furthermore, does the expression of the 'chemist'—'This particular material group of matter is the foundation for the eye'—mean? How can the tendency to development which makes itself quite evident in a fertilized egg be brought under a structural formula? What, for the chemist, is 'growth' and evolutional energy?[1] Is a chemical formula imaginable which can express 'inheritance'?[2] What formula has conscience, and what structure and organization can present it graphically? All that, however, belongs to 'life,' and should be explained.

[1] Reinke: *Die Welt als Tat*, p. 293.

[2] Bunge well expresses this (*Lehrbuch d. Physiol. d. Menschen*, II): 'But we can perfect the aids to investigation! We can increase the powers of the microscope! The cell, which appears structureless to-day, will show a structure to-morrow. . . . And the nucleus also is no longer structureless. . . . But—a complex structure is no explanation: it is a new enigma. How has this complex structure originated? Will it perfectly solve the great enigma, the greatest of all—the enigma of inheritance—the inheritance through a small cell?'

(ii) *The attempts to present the process of evolution in a concrete form demonstrate the impossibility of spontaneous generation.*

(*a*) The attempt of Naegeli[1] to render comprehensible the origin of the simplest organisms from inorganic matter must be regarded as a complete failure. The 'being originating from spontaneous generation' must, according to him, 'be in the first place perfectly simple' 'without external form and without internal members,' pure albumen, which then nourishes itself. That 'scarcely merits the name of an organism, but it may be the commencement of a series which leads to an organism.' 'Growth and reproduction gradually acquire by inner relations greater definition,' etc. In this way 'gradually all qualities of the monad are newly generated.'

Shortly stated, the entire allegation is to the effect that a cell, as it now is, is first analysed and then again brought together piece by piece, whereby nourishment and reproduction gradually come in as firmly established peculiarities. The first really living being that we know of (according to Naegeli the 'monad') 'must, in the organized arrangement of its parts, be already far advanced and therefore have a long series of ancestors behind it.' Certainly! That, however, which existed before the 'monad' was the purely hypothetical 'probien' which, if they lived, must also have had that organic

[1] Naegeli: *Theorie d. Abstammungslehre*, pp. 83, 86.

arrangement and a long line of ancestors, and, if they had not these, did not live at all, but were 'drops of albumen of the most perfect simplicity.'[1]

(*b*) The hypothesis recently put forward by Mereschkowsky,[2] of the two kinds of plasma, only concerns us here in so far that it also tries to explain spontaneous generation.

As claims, 'which unavoidably must be made for the first organisms,' Mereschkowsky cites minute submicroscopic size, absence of organization (?), capacity of standing high temperatures, of living without oxygen, of forming albumen and carbo-hydrates (starch and sugar) out of inorganic matter, and great resistance against strong salt solutions and poisons. All this we see, so he continues, demonstrated in the bacteria. Therefore they must have been the first organisms. It may be that the bacteria were the first organisms, but from Mereschkowsky's statement that by no means follows. But since these original beings were organisms —the visible organism we will willingly accept as non-present—and sturdily maintained, nourished, and reproduced themselves, they were consequently perfect living beings with an organization in the biological sense. How, however, did these living organisms arise from albumen—particularly the power of reproduction,

[1] Naegeli: *Theorie d. Abstammungslehre*, p. 86.

[2] Prof. Dr. C. Mereschkowsky (*Biol. Zentralbl.*, 1910, p. 278: *Theorie der zwei Plasmaarten als Grundlage der Symbiogenesis, einer neuen Lehre von der Entstehung der Organismen*) explains the origin of all higher organisms by symbiosis of mycoplasma and amœboplasma. In a general way, as this doctrine is put forward, it is not to be taken seriously.

this stubborn tendency to self-preservation ? These arose, according to Mereschkowsky, in the following way: As gradually the conditions for the formation of new mycoplasma became more unfavourable 'the albumen began to decompose, to decay, and it could no longer build anew. By virtue of this there consequently disappeared the conditions for the formation of living mycoplasma, and the further development of life could only proceed on the principle of *omne vivum e vivo.* There arose at once one of the main differentiating peculiarities of life—the capacity of reproduction, i.e. of commencing new beings from living parts of the old ones, since only such particles of albumen could flourish which possessed this faculty, and had such not been produced, then there would have been no life on the earth.' [1]

Thus, because albumen commenced to decay, the principle of *omne vivum e vivo* had to arise, since, had it not arisen, there would have been no life. Is that a scientific explanation of spontaneous generation ?

Such attempts render J. Reinke's words comprehensible which he wrote of Naegeli: 'Thereby I think I have shown . . . that the grounds put forward by him for the occurrence of spontaneous generation can hold water so little that they make the spontaneous origin of organisms appear as absolutely unthinkable.' [2]

We cannot, therefore, see how O. Hertwig [3] can term

[1] *Biol. Zentralbl.*, 1910, p. 362. [2] Reinke: *Die Welt als Tat,* p. 337.
[3] *Allgemeine Biologie,* p. 270.

the acceptance of spontaneous generation—for which he himself can find no grounds at all—a philosophical need.

With the reason for this Häckel should provide him; he says: 'We must regard this hypothesis as the immediate consequence and the necessary completion of the generally accepted theory of the earth's formation of Kant and Laplace; and we find therein, in the totality of natural phenomena, such a compelling logical necessity that we must therefore regard this deduction, which to many appears a very bold one, as incontrovertible.'

The theory of the earth's formation, of Kant and Laplace, has as 'immediate consequence' that the life on the earth once began and was not always there, and that is all; regarding spontaneous generation the theory states nothing.

Regarding the 'totality of the natural phenomena' which here comes into question, we can quote, also according to Hertwig's own investigations, the following sentence: 'In the "totality of the natural phenomena" which here actually come into consideration, we find such a compulsive logical necessity that we must regard the denial of spontaneous generation as incontrovertible.'

To these 'natural phenomena' belongs in the first place the experimentally determined fact, accepted by all biologists ('axiom' as it is often called), that life now only arises from the living.

To this belongs the 'impossibility' of regarding

vital activity as a chemico-physical process even of the most complicated kind, or of regarding the organism as merely a machine composed of material parts held together by material forces.[1]

It is important to note also the 'phenomenon' that all attempts have entirely failed to render spontaneous generation more comprehensible by means of the natural forces known to us, even with the aid of quite 'peculiar' conditions. Lord Kelvin speaks of physical 'hocus-pocus.' Helmholz and Arrhenius seize upon the boldest hypotheses in order to avoid 'spontaneous generation.' Reinke and E. v. Hartmann find that precisely these 'attempts' show the inconceivability of it. O. Hertwig confesses that, 'owing to the present position of natural science, the investigator has certainly no better prospect of results in obtaining living from non-living material than Wagner, in Goethe's "Faust," of brewing a "homunculus" out of a retort.' This firm conviction is therefore certainly based upon the totality of the natural phenomena.

Nor do we avoid the 'miracle' (and certainly this time a true one) by accepting 'spontaneous generation.' Every living organism shows to-day the persistent striving to combat, by constant contrivance (*Mauserung*), the ageing and hardening tendency of its material basis,[2] by fresh formation of highly complex and unstable

[1] *Allgemeine Biologie*, p. 159. 'Therefore [by reason of the "essential" differences between mechanical action and vital activity], it is an entirely vain endeavour to imagine that an organism can be understood on the principles of mechanics.'—Hertwig.

[2] E. v. Hartmann: *Das Problem des Lebens*, p. 204.

(*labile*) chemical combinations. If, therefore, the organisms arose from inorganic matter, then they must be struggling constantly with ever-increasing energy against their own nature, which compels them to form the most stable connections possible. 'Every smallest increase in the complication of the chemical combination and in the instability (*labilität*) is contrary to natural laws, in so far as mechanical conditions are not accepted, as they cannot arise by themselves in conformity with inorganic laws.'[1] If, despite this, 'spontaneous generation' has occurred, then there has happened a 'miracle' in the sense of modern natural science—i.e. a breach of Nature's laws.[2]

(*c*) Into the attempts to explain the origin of life by comparison with liquid crystals and by all sorts of experiments with gelatine and artificial trees, we shall not enter. If a liquid crystal actually lives, why is it not left to its fate, so that it may, like bacteria, go further and reproduce itself? That two 'liquid' crystals run together is very natural; but it is naive to speak of copulation in the sense of a melting of cells. If, perhaps, reduction division occurs, have the crystals, by complicated and wonderfully purposeful processes, become desirous of and capable of fertilization?

The 'artificial plants' constitute certainly a very amusing and interesting experiment, similar to the well-known 'Pharaoh's serpent' in chemistry; but if

[1] E. v. Hartmann: *Das Problem des Lebens*, p. 192.

[2] Reinke says somewhere that if a spontaneous generation be accepted it might also be maintained that water formerly flowed uphill and not down.

it is desired to originate life thereby, the words of Lord Kelvin, the great physicist, apply: 'No hocus-pocus of electricity or physics could make a human cell.'[1]

Rhumbler also—certainly a reliable and creditable investigator—says of the pseudo organisms: 'The whole of the pseudo organisms resemble the true ones, even under the most favourable conditions, only in the sense of similar or like configuration and in similar or like distribution of the aggregate parts. Thereby the inorganic substance does not approach the peculiar existence of the organic by a hair's breadth, any more than does any other physiological model its living original—or let us say, for better exemplification, than a model of a heart consisting of a rubber bag and the necessary pumping apparatus would approach a living heart whose beating only is represented by the model.'[2]

§ 2. *We are not justified in bringing animals and plants into genetic connection.*

By animals we understand, as a preliminary, organisms like Mammalia, Birds, Fishes, Worms; by plants, Trees, Ferns, Mosses. All that falls under the general category of 'life,' as we have already stated, embraces these beings. But a dog behaves in many vital ways differently from a fruit tree—for instance, in the activity which he exhibits in seeking food or the other sex. In it we observe vital expressions, which are similarly

[1] See the excellent monograph, *An des Lebens Schwelle*, Prof. E. Klein, Luxemburg, 1909, p. 19.

[2] *Ibid.* p. 16.

executed as are, by man, those actions which are consciously done, and which he wills to do in order to attain a certain object which he recognizes as, and feels to be, desirable. The necessary movements in such a case—movements of the whole body from place to place, or movements of certain parts (the hand for instance)—are then so regulated as the external impulses in the particular case require. The behaviour of a hungry dog who rushes towards a piece of meat may serve as an example. The meat is obviously his recognized object, since thereto he directs his staring eyes, as we men are accustomed to do when we have observed something. Thither he wishes to go, since he resists if he is restrained; he drags at the chain, always in the direction towards the meat, if he be withheld; he feels strongly this striving within himself, since he howls and whines if he be hindered in reaching the desired object. In short, he behaves like a man, who in an analogous case allows himself to be guided only by his sensual appreciation and impulses. Since, therefore, the dog shows all expressions of a conscious sensual action, we must concede to him also the pre-supposition implying sensual appreciation and impulse; otherwise we have an insoluble enigma before us.

This is so because unconscious purposeful movements either do not come, in such cases, at all into consideration—as for instance the automatic ones—or they are not sufficient—as with the reflex ones. 'The reflex respond to each impulse with machine-like regularity, always in the same way, without the slightest deviation :

their action is monotonous.'[1] The dog, however, who strives after his food takes account of the hindrances as they present themselves—he acts individually to a definite end ; for that there suffices no organism of machine-like arrangements, and therefore there is no reflex activity.

All organisms, however, do not behave like the dog ; there are many to which none of the three criteria apply which are used in comparative physiology in the search for conscious (' psychic ') vital expressions, and which require the acceptance of true recognition and voluntary power of striving. There are lacking, namely, in the first place the easily observable external signs of spontaneous (voluntary) movements seen in the larger organisms ; a use or a necessity for the vital impulse appears, furthermore, to be excluded, since reflex and other arrangements in the construction secure the undisturbed performance of all the vital functions ; finally the ' organs,' which may resemble animal ones, show themselves in the clearest way to be arrangements for the reception of definite external ' impulses.' The first two criteria—the lack of voluntary motion and the absence of any necessity for the vital impulse—can be deduced directly from the absence of sense faculties in ' plants ' ; the third—the possession of special ' organs '—proves nothing, if it be not previously declared what has to be proved. Since, however, there are frequent references to the new investigations,

[1] F. Lucas : *Psychologie der niedersten Tiere,* Vienna and Leipzig, 1905, p. 11.

particularly by Haberlandt, into the 'sense organs of plants,' we will go more closely into the matter.

What, in the opinion of modern biologists—and particularly also of Haberlandt—is 'excitability' (*Reizbarkeit*) and what does its presence imply for the organism which possesses it ?

What signifies the construction of particular 'organs' for the reception of special stimuli, such as that of light, mechanical pressure, breakage, gravity, etc., for the possession of actual sense faculties—recognition and voluntary power of effort ?

From what has already been said regarding the relation of the organisms to inorganic matter and the general forces of nature, such as light, warmth, gravity, there is a double deduction.

Firstly, the organism needs, if it will be active, an influx of energy from outside ; secondly, it uses this energy for its own purposes, such as building material for instance, the formation of digestive (assimilating) tissues, of conducting systems for the nutritive matter through the whole body, of germ cells from which new individuals of its kind result, etc.

Oxygen, hydrogen, carbonic acid, sulphur, phosphorus, lime, and the other elements which exist in the organism, also light, energy, electricity, gravity, have quite obviously in themselves no tendency to build tubes for water conduction or blood circulation, or a leaf which breathes and assimilates, or an egg and seed-cell of a fir tree. If, despite this, the chemical material is necessary so that the activity of the organism may be excited and rendered possible, then can they only be regarded as 'impulsions,' as building material and sources of energy; they form, as is very significantly expressed, 'stimuli'—i.e. impulses and material for carrying on the life of the organism. The capacity of the living body to respond to an external impulse with vital expressions, such as growth, formation of fruit, or attractive or repulsive movements, is called 'excitability' (*Reizbarkeit*).

'Excitability,' taken in this sense, signifies thus as much as vital capability in general and is the peculiar mode of reaction of all organisms in response to external influences in contrast to the behaviour of inorganic bodies under like influences. We are not alone in this conception: it is put forward and established by very eminent biologists, such as Pfeffer, Strasburger, O. Hertwig, Sachs, and others.

In the 'Lehrbuch der Botanik' of Strasburger, Noll, Schenk, and Schimper[1] so extensively used in the German high schools, there is the following definition of 'excitability': 'It is shown thereby that external or internal impulses given to the living organism act as dissociating stimuli and induce activities which it effects with means over which it has control or which it is capable of obtaining, and in a manner determined by its construction and by its needs. Even in the smallest and simplest organism of which we know, the vital processes depend on such stimuli.'

[1] *Lehrbuch der Botanik für Hochschulen*, Jena, 1900, p. 4.

According to Pfeffer[1] the stimuli are associated with the whole of the vital action, and there is perhaps no single action 'in which these do not and must not play a rôle . . . they are thus a general quality of all living substance.'

Haberlandt calls 'all organisms, animals and plants, excitable.'[2] Then follows almost word for word the definition given above by Strasburger. Similar opinions were already held by Treviranus; Haberlandt says of him,[3] 'that he had correctly grasped the signification of vital excitation.' Treviranus, however, attributes the whole vital process to stimuli.

The term 'excitability' may, however, be more narrowly defined, as is done by Haberlandt in his work 'Sinnesorgane im Pflanzenreich.' He discusses therein the special arrangements of many plants whereby 'the sudden deformation of the sensitive protoplasm which is essential to the "excitation" becomes particularly easy and marked. This is also the most general building principle of this apparatus.'[4]

Many plants have thus—and this Haberlandt has described in a masterly fashion and determined by experiment—special apparatus for receiving definite stimuli—for instance, mechanical contacts. Such

[1] In *Pflanzenphysiologie*, I, Leipzig, 1897, p. 10, Pfeffer objects 'emphatically' to the assumption that only certain striking phenomena of motion were due to excitation, as, for instance, the sudden movement of the sensitive plant *Mimosa pudica* (p. 11).

[2] G. Haberlandt: *Physiologische Pflanzenanatomie*, Leipzig, 1909, p. 520.

[3] *Sinnesorgane im Pflanzenreich zur perzeption mechanischer Reize*, Leipzig, 1906, p. 4.

[4] Parenthesized in the original.

apparatus are particularly favourably placed in the organism, and their entire anatomical structure shows clearly that their object is that the external influences shall act vigorously and directly upon the right spot. Does this afford any proof that the actions, which are produced by such 'organs,' are accompanied by consciousness?

Quite certainly this is not the case. According to Haberlandt himself a similar capacity for perception of mechanical excitation exists in other plants, but 'diffused'—for instance in all the cells of a leaf or leaf tissue or stalk.[1] The development of particularly localized apparatus with the exclusive function of responding to excitation increases, it is true, the general 'faculty of perception' and can be adapted to special purposes—as, for instance, sudden and powerful movements, but it alters the nature of the excitability absolutely not at all.

So long, therefore, as it is not shown that 'excitability' implies quite general 'psychical' qualities, there is, by those organs, nothing gained at all for the acceptance of conscious vital phenomena in plants. A careful study of the illustrations enables us also to recognize without difficulty the typical form of reflex mechanism, not only in the organs for mechanical stimuli but also those of gravity and light.

Furthermore Haberlandt quite emphatically remarks that for him the 'psychical' side of the sensitiveness 'is an accompanying or parallel phenomenon outside

[1] *Physiologische Pflanzenanatomie*, p. 520.

the scope of his investigations.' [?] For us that is not the case, since here it is not a question whether the plants really have organs for the reception of stimuli, but whether they have sense organs. We can and must separate these questions, since with us men there occur purposeful excitations which happen unconsciously (for instance reflex action), and such as we voluntarily produce, with consciousness due to an internal or external stimulus.[1] Among the animals also we have both kinds of movements; to the plants therefore we must apply the psychological criteria in order to settle the question. That, however, was not the case with Haberlandt; his text and his illustrations permit, moreover, of only the one deduction—that we have to deal with indubitable mechanisms, with reflex actions.

Conclusion.—There are, therefore, organisms which show clear expressions of conscious vital action, of sensible recognition (sensations and feelings), and striving power, and others in which such expressions are never observed, not even in the most imperfect state. The first we call 'animals,' the second 'plants.' Consciousness we cannot, however, regard as something of secondary importance, since 'the entry of the conscience into the series of vital phenomena must not appear to us as an incomprehensible miracle but as something natural and easily comprehensible, and that is the case if consciousness, from the very commencement, already at its first appearance has a definite task to fulfil in the service of the whole organism like every other

[1] See Wasmann: *Instinkt und Intelligenz im Tierreich*, 1905, p. 6.

vital phenomenon.'[1] The entire vital activity depends, then, in animals upon their sense faculties,[2] since reflex action alone would not suffice either for obtaining food, or for protection against attack, or for reproduction. In plants, on the contrary, it is positively provided for that they have always their nutrition surrounding them—air, water and salts in solution—and that the pollen reaches the female flower by self-pollination or with the aid of the wind or even of insects specially fitted for the office. Consciousness is here positively superfluous.

We came earlier to the conclusion that in each organism a single principle is to be accepted which stands above matter and utilizes its materials and energies unitedly, since inorganic material never shows the trace of a tendency to collect into mutually co-operating systems with a purposeful division of the vital functions. Now we have learnt to know of organisms, the animals, into which a new factor in vital activity is purposefully interposed—i.e. sensible recognition and voluntary power of action. That is, that there is a principle, which in the dog, for instance, leads and directs the life, and possesses also the sense faculties, since, exactly like the other faculties, they are utilized to one and the same vital end. Thus the entire 'psyche' of the animal is another than that of the plants, which display no consciousness. Should, therefore, a plant

[1] F. Lucas: *Psychologie der niedersten Tiere*, Vienna and Leipzig, 1905, p. 18.

[2] See Wasmann: *Die psychischen Fahigkeiten der Ameisen*, 1909, p. 5.

become an animal, the whole nature of its principle must be transformed, and only when this transformation is complete can consciousness show itself for the first time. A thing 'a,' however, which can only become a thing 'b' by a total alteration of its whole being, absolutely excludes all connection with 'b' by gradual evolution; because though evolution may effect development of a basis or the improvement of a completed object if the necessary evolutional tendency leads thereto, it cannot effect new and higher modes of existence. Animals and plants cannot therefore be brought into genetic connection; the question of the origin of the animals from plants forms no problem of the hypothesis of evolution.

§ 3. *We are not justified, in the present state of our knowledge, in bringing the families and classes of the animal and plant worlds into genetic connection.*

(1) *Introductory remarks on the systematic treatment of animals and plants.*

The systematic division of the animal kingdom has, since Cuvier, been effected mainly according to two points of view[1]—according to their affinity to a definite type (form of construction), and according to the height of their organization. The type arises from the mutual relative positions of the organs in the organism and the symmetry of the whole; the height

[1] See, e.g., B. R. Hertwig: *Lehrbuch der Zoologie*, Jena, 1910, p. 104.

of the organization from the degree of perfection—i.e. from the greater or less strictly executed distribution of the separate vital functions to special tissues, i.e. organs.

The 'type' determines by itself alone nothing as regards the perfection of the animal; the degree of differentiation in separate tissues and organs yields, however, an objective criterion—i.e. one based on reality—of the height of organization. That applies particularly also to the differentiation of the nervous system.[1] According to the 'plan of construction' and the height of organization there is effected also at present the systematization of the animals and plants into a few well-defined groups.

As a rule the zoological textbooks divide the animal kingdom into seven, nine, and even more families (*Stamme*). R. Hertwig holds to seven in his well-known 'Lehrbuch der Zoologie.' E. Selenka[2] adopts ten and distinguishes them as follows:

I. Monocellular	Protozoa (1)
II. Leaf animals (Metazoa) with cellular differentiated tissues and organs:	
1. No bodily cavity as blood reservoir; Coelenterata:	
(*a*) Bodies non-symmetrical	Spongiæ (Sponges) (2)

[1] Recently here and there the biologists have been denied all right to speak of various grades of perfection or heights of organization, or, at the best, they are yielded to only for practical reasons (by reason of the division of matter). See B. Franz in *Biol. Zentralblatt*, 1911, p. 1: 'What is a higher organism?' We shall refer again to this article.

[2] *Zoologisches Taschenbuch*, I, Lepizig, 1897, p. 1. For quick information the two volumes of this handbook are much to be recommended; they will always provide a clear and short grounding in the systematic arrangement.

(*b*) Bodies four to six-rayed; possess stinging cells	Cnidaria (3) (Polyps and Medusæ).
2. Body cavities (at least as a split); Coelomata: (*a*) Body five-rayed (symmetrical); body wall strengthened with a lime skeleton	Echinodermata (4) (Sea Urchins, Starfish)
(*b*) Body laterally symmetrical, bilateral:	
(*α*) Central nervous system is not a dorsal tube; no internal axial skeleton:	
A. Body without limbs, or, if limbed, without divided extremities	Vermes (Worms) (5)
B. Bodies without limbs, with three pairs main ganglia (nerve cell agglomerations) with foot, mantle, and shell	Mollusca (6) (Soft animals, Snails, Squids, etc.)
C. Bodies limbed with divided extremities	Arthropoda (7) (Insects, Spiders, Crabs)
(*β*) Central nervous system in spine; bodies limbed:	
A. As axial skeleton only a chorda dorsalis (dorsal chord, a soft rod); Chordata Limbs and Chorda confined to hinder parts of body	Tunicata (8) (Mantelthiere, Seescheiden).
Limbs extend throughout the body	Leptocardii (9) (e.g. Amphioxus, the Lancelet fish)
B. With the Chorda is associated the spinal column and the skull	Vertebrata (10) (the Vertebrates)

For the tabulation of subdivisions (classes of the family) no generally available criteria can be given. A textbook of geology should be read in that connection; as concerns systematic arrangements Selenka's 'Zoologisches Taschenbuch' is the best.

We will now give an example from the last-named work.[1] Many of the characters described must be simply taken as 'given.' They contribute much to the total habit, but we cannot speak of higher or lower forms.

Typical Differences between Reptiles and Mammalia.

Reptiles.	Mammalia.
Skin scaly, offering no protection against cold; temperature changeful (*Poikilotherm*).	Skin protected from cold by hairy coat or cushion of fat; temperature even and high.
Larger and smaller circulation, mostly and imperfectly separated; slow assimilation.	Perfectly separated; active assimilation.
Lungs sack-like, of small superficies reaching to the abdomen; chest small.	Lungs alveolar, with large surface confined by diaphragm (breathing muscle) in thorax; chest broad.
Mainly dependent on animal food and warm climate. The ovum is developed by aerial temperature.	Manifold provision, hence distribution more extended. The ovum is hatched by maternal warmth.
Teeth alike; persistent change of teeth (polyphyodont). •Vegetarians are monophyodont or anodont (Tortoises).	Teeth dissimilar; stomach relieved by grinding of food; jaw typically diphyodont.

[1] Selenka: *Zoologisches Taschenbuch*, I, p. 190.

REPTILES.	MAMMALIA.
Front of brain small, olfactory centre well developed.	Front of brain large; all organs of sensation possess higher sensitive centres in the brain; gradual progress in intelligence (?).
One olfactory muscle.	Three or more olfactory glands.
Neck immoveable; one condyle (swelling behind head).	Neck moveable; two condyles.
Oviparous; eggs with shells and with abundant yolk.	Viviparous; eggs with little yolk or none. Ovum is constantly nourished by maternal secretions (uterine mucus, uterine milk, blood serum and, later, milk).
Allantois (an embryonal formation) is the urine bladder and breathing organ.	Allantois is only the bladder (many marsupials) or also the bearer of the embryonal placental vessels.
Greater abundance of forms in the Permian formation and until those of the Jura and the Chalk.	Greater abundance of forms in the recent Tertiary formation.

The orders in the class of the Mammalia are determined essentially on the basis of different construction of the extremities and of the jaw.

The plants are generally divided into five series or families. This is done mainly according to the degree of differentiation in the first place of the greater groups of tissue—the roots, stem, leaf and flower—then according to the greater or less perfect distribution of work in the organ formation and the purposeful constitution of the separate tissues (conduction, assimilation, protection, etc.), and according to the reproductive arrangements.

Warnung[1] establishes five series.

1. Thallophyta or Sessile Plants.—One or more celled plants of simple construction, almost always without limbs in the root, stem, and leaves (such an undivided plant body is styled Thallus), and always without vascular bundles[2]—Algæ, Fungi, and Lichens.

FIG. 25.—A Moss—*Hypnum Purum*. k, the spore capsule with its stalk, seated on the 'moss plantlet' and produced from a fertilized ovum.

2. Bryophyta or Mosses.—Small plants with thallus or leaved stalks, but without true roots[3] and with vascular bundles. The group of Mosses is sharply differentiated from the following groups by the peculiarity of its reproductive changes.

The green moss plant—that is what the layman generally understands by 'moss'—produces in separate receptacles both ova and motile antherozoids. The ova receptacles are called archegonia, those of seed cells antheridia.

[1] *Handbuch der systematischen Botanik*, German edition, Berlin, 1902, p. 1.

[2] Vascular bundles are essentially bundles of conducting tubes for the transport of water and the dissolved earthy salts contained therein to the leaves and other organs.

[3] 'True' roots—i.e. roots with varied tissues. The Mosses have only 'Rhizoids'—i.e. threads formed of like rows of cells or even only of undivided protoplasmic tubes. Rhizoids are therefore imperfect—'pseudo'—roots.

From the fertilized ovum the green-leaved moss plant is not immediately produced, but it forms in the first place another 'plantlet'—the sporogonium (Fig. 25). When complete it consists of a stalk (seta) which carries a capsule at its tip. The stalk is sunk within the mother plant, without, however, an actual connection by growth existing. In the capsule there are finally numerous spores which are newly formed out of an 'indifferent' tissue.

By the bursting of the capsule the spores are hurled out, and only from these are there again developed the green 'moss plants.'

In regular succession there thus arise individuals which produce the sexual cells proper—viz. ova which need fertilization, and seminal threads (antherozoids in the plants concerned), and form others,[1] the spores [2] or germ cells, protected by a firm skin, which never require to be fertilized, at least not with seminal threads, in order to develop. This regular succession, one after the other, of a sexual and asexual generation, is termed 'alternation of generations.'

[1] Is the Sporogon an actual 'individual'—i.e. a new actual plantlet—or an organ of the mother plant for the formation of reproductive cells? It is very difficult to decide the question, but for our description it has no importance, since we here consider only the actual differences which can justify us in determining separated types. For the actual plant (not organ) nature there speak the formation of various tissues with separate vital functions, assimilation tissues, breathing apparatus (stomata), and the circumstance that the germ cells (spores) are newly developed from indifferent material. The spores are therefore not the immediate product of the self-dividing fertilized ovum cell.

[2] It would go too far to enter exactly into the definition of the 'spore.' The definition given above in any case suits the plant spores generally.

3. Pteridophytes or Ferns.—Plants with stalk, leaves, and true roots; vascular bundles without 'true' vessels.[1] The alternation of generation is quite different from that of the Mosses. The green plants (which the layman only knows as the Fern) carry spore-containing vessels (in which there is no ovum as

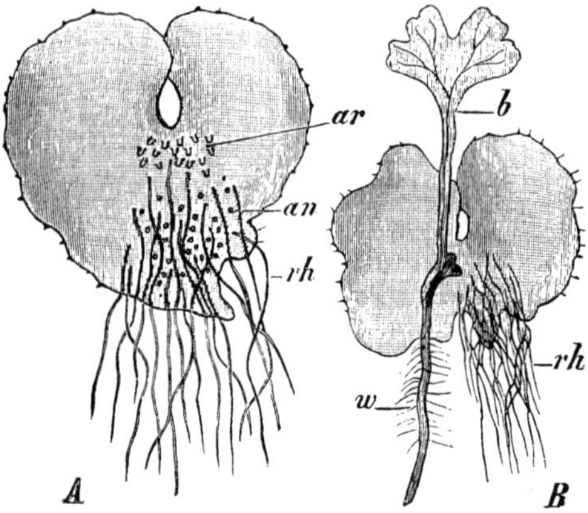

FIG. 26.—NEPHRODIUM *filix mas*. *A*—Prothallium from the under-side with archegonia *ar*; antheridia *an*; root-hairs *rh*. *B*—Prothallus with young fern produced from a fertilized ovum. *b*, fern leaf; *w*, root of same. Mag. about 8 diam. (*After Strasburger*.)

with the Mosses). From the spores scattered on the ground there are developed so-called prothalli, in which true ova and seminal threads are formed. It is only the fertilized ovum which reproduces the green fern plant directly (Fig. 26).

[1] 'True' vessels (trachei) = conducting tubes in which the transverse walls of the rows of cells which form such a tube become absorbed. By this means the conduction of water is certainly facilitated and hastened. In the 'Tracheids' the transverse walls of the separate cells are retained. That is the main difference.

Two sorts of spores occur—microspores which form prothalli with seed-vessels, antheridia, and macrospores which produce such with embryo ova (archegonia).

Four classes of this series are determined: Ferns (Filicinæ), Equisetums (Equisetæ), Lycopods (Club-mosses), Water Ferns (Hydropteridæ).

4. Gymnospermæ or Naked-seeded Plants.—Woody plants with separated male and female flowers. In the stamens are produced multicellular pollen grains which are carried by the wind direct to the more or less obvious (naked-seeded) female embryo seeds. Several embryo ova are found in the embryo seed.

The first classes of this series are: Cycadinæ (Sago or Fern Palms), Ginkgoinæ (a still living species is *Ginkgo biloba*), Coniferæ (Fir Trees), Gnetinæ (a small exotic group).

5. Angiospermæ or plants with covered seeds.—The flowers often contain both male and female organs at the same time; more rarely these are borne in separate flowers or even separate plants. The pollen grains are only two-celled; fertilization can only occur when the pollen grains, on germination on the stigmatic surface, send a tube down through the entire stalk till it reaches the entirely enclosed embryo ova in the seed vessel. Each embryo only contains one ovum.

The chief divisions are: Monocotyles and Dicotyles. The series 4 and 5 are distinguished from the other three as flowering or seed-bearing plants. A flower is produced at the terminal point of a shoot and consists exclusively of reproductive organs. The leaf formation is transformed either into fruit or pollen-bearing leaves, or into protective or attractive means in connection with the fertilization. 'Seed plants' they are called because it is through the seed that the formation of new individuals, separated from the mother plants, is effected. A seed is a multicellular body, which, when it leaves the mother plant, is already differentiated (a grain of wheat, an apple pip).

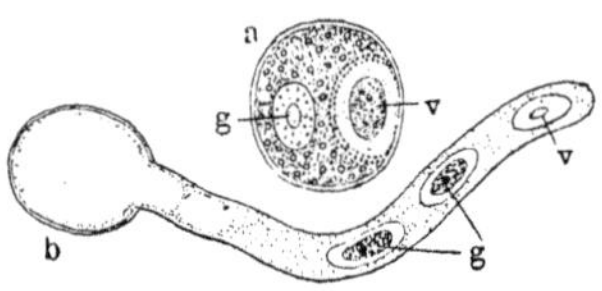

FIG. 27.—a, Ripe pollen grain of an Angiosperm ; b, 'germinating' pollen grain ; g, generating germ ; v, vegetative germ.

It has been possible to establish the existence of an alternation of generations in the Gymnosperms and Angiosperms, but certainly of a very debased kind. A series of indications point to the pollen grains, and the so-called embryo sac in the bud germ, as being 'spores,' since they, in the first place, without fertilization, form a sort of prothallus (second individual ?) in which first arises the fertilizing cell proper (sexual cell or, better, sexual germ). The prothallus possesses, it is true, in some Gymnosperms only three small cells, or even only one in the Angiosperms—the so-called vegetative cell (Fig. 27). The macrospores produce, on the other hand, in the Gymnosperms a particular

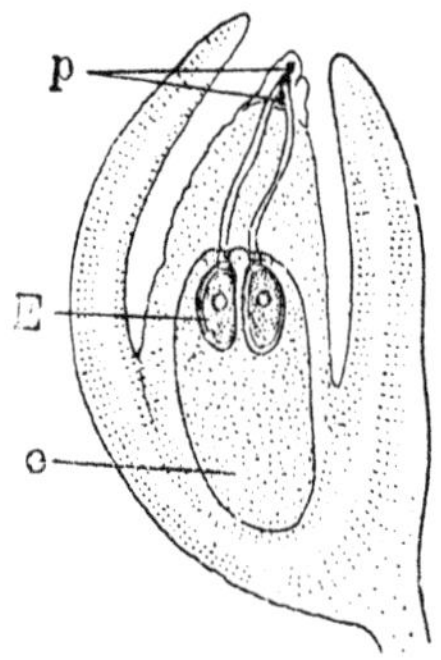

FIG. 28.—SECTION THROUGH THE EMBRYO SEED OF A GYMNOSPERM (*Picea vulgaris*).
E, ovum ; e, embryo sac (Endosperm) = prothallus ; p, two pollen grains which send tubes down to the embryo ova.

tissue in addition (endosperm) in which several ovum cells can be deposited (Fig. 28). In the Angiosperms there still remain three cells, which may be regarded as a prothallus (the so-called three antipodes). There is never more than one ovum formed (Fig. 29).

The determination of classes within these series is effected mostly according to the construction and the position of the spores —or the ovum and seed vessels, the formation and arrangement of the leaves, the small or tree-like constitution of the stem (stalk), etc. From all this there results a habit of growth which is fairly characteristic but can only be sufficiently known by seeing the plants.

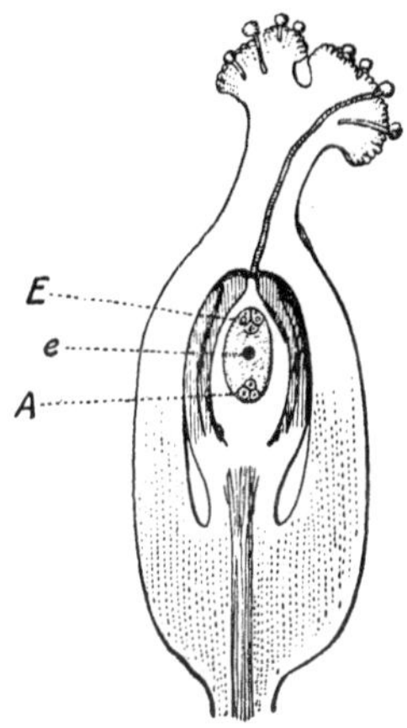

FIG. 29.—*E*, Fruit bud of an Angiosperm (*Polygonum convolvulus*) during fertilization. Pollen grains and pollen tubes as above. *e*, embryo sac; *E*, the one ovum cell (the two others do not become fertilized). *A*, the three antipodes (=prothallus).
(*After Strasburger.*)

As an example we will describe the classes of the Pteridophytes (after Warnung).

Class 1: True Ferns (Filicinæ) and Hydropteridæ (Water Ferns).

(*a*) Habit mostly herbaceous; leaves alternate, large in comparison with the stem and highly developed (feathery); when young, circinately rolled (mostly).

(*b*) Peculiarities of the reproductive organs: the spore-cases are situated on the edge or on the back of ordinary leaves; only in some cases are the fruitful (spore-bearing) leaves specially changed. The fertile leaves are not confined to definite parts of the caudex and do not limit its growth.

The 'Water Ferns' are small plants with a horizontally growing stem which creeps on the ground or floats on the water. They are heterosporous—i.e. they produce in separate vessels micro- and macrospores; the true Ferns are homosporous, with only one kind of spore.

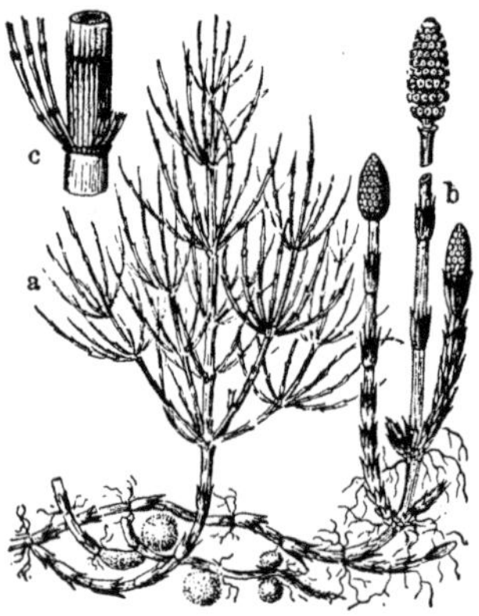

FIG. 30.—HABIT OF AN EQUISETUM (*Equisetum arvense*). a, barren shoot; b, fertile shoot; c, portion of stem.

Class 2: Equisetinæ (Horsetails).

(*a*) Herbaceous (all the present forms); the leaves relatively small, undivided, arranged in whorls, and branched (Fig. 30).

(*b*) The spore capsules are situated on peculiarly metamorphosed leaves which are conjoined into a single bloom (so called) which terminates the growth of the shoot. The branches are arranged in whorls—i.e. they spring at even distances and all in the same plane on the stem.

Class 3: Lycopodinæ (Club-mosses).

(*a*) Herbaceous (all the present ones); the leaves very small and simply constructed (almost scale-like).

(*b*) The sporangia (spore vessels) are situated singly on the base of the upper side of the leaf, or in the leaf axis, or above the axis on the stem itself.

The 'fertile' leaves, which are united in special clusters at the tips of the shoots (stalks) and hence terminate the growth (Fig. 31), are very often differently formed to the barren ones.

(2) *General results of systematic classification compared with those of palæontology.*

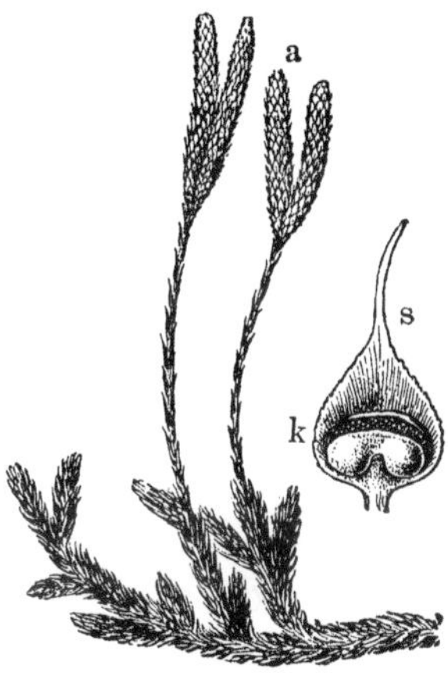

FIG. 31.—HABIT OF A CLUB-MOSS (*Lycopodium clavatum*). a, so-called 'bloom' = cluster of spore vessels; k, separate sporangium burst open; s, leaflet (scale).

From the above it is clear that *a priori* we cannot know whether our systematic categories are really primary and do not signify forms of animals and plants which are reducible one from another, i.e. true types in the sense of Cuvier. Both elements which present themselves in the definition of a 'type'—plan of construction (symmetry, habit as a whole), and the degree of differentiation in tissues and organs for the general vital activities—can in themselves be subject to variation. It may therefore happen that family and class characters may have arisen out of other forms.

It is therefore from the outset not excluded that, for instance, all plants which have been assembled under the classification of Ferns are only various modifications of one and the same form of growth and have arisen through the development and evolution of one original form.

The possibility that Ferns, Equisetæ, and Club-mosses might be traced back to a common starting-point cannot be *a priori* disproved.

But one circumstance must be expressly emphasized. The differences which exist between the classes and the families of animals and plants are the first in the organic kingdom which may be bridged over by evolution: the question whether and how the associated forms now in existence can be regarded as the results of developmental (evolutionary) process is the first problem of an evolutional hypothesis. The evolution must not here, in this question, be regarded as in any way put forward as a fact: that would be quite inadmissible.

No one has so far maintained that, on closer study of the relations between the inorganic and organisms, or between animals and plants, positive starting-points for the acceptance of an evolution have been found. Spontaneous generation is put forward by investigators exclusively as a postulate, as we have seen; the transformation of a plant into an animal is little discussed, and it is simply impossible to explain it by 'evolution.'

Consequently the testing of the connection of the greater systematic groups among themselves is of fundamental importance for the evolution question, since here for the first time a genetic connection is not from the outset excluded.

From the results of the systematic classification

of the present organisms, compared with those of palæontology which we have discussed above, there are deduced, as it appears necessarily, the following conclusions:

1. The generally accepted and usually corresponding separation of all the recent organisms into a number of groups shows that there are really certain gradings existent. We are in the great majority of cases not long in doubt whether a plant, for instance, belongs to the Ferns or Equisetums or Angiosperms.

If difficulties arise, they do so almost always on account of peculiarities, which we may justifiably consider as due to secondary 'adaptive phenomena' (specialization or regression). The study of the reproductive relations, with which in most cases are associated other characteristic features, finally determines the systematic position.

2. All certain results of palæontology indicate that the still surviving higher systematic categories were retained also during the geological periods, so that we can class the greater part of the fossil forms with the recent ones in the same system.[1]

Of clear transitions between Pteridophytes, for instance, and Gymnosperms nothing can be said; what

[1] To class together fossil and recent forms as equivalent and closely associated orders, families, etc., is the endeavour of all the newer text-books—for instance, the great joint work of Engler and Prantl: *Die natürlichen Pflanzenfamilien*. See also Lotsy: *Vorträge über botanische Stammesgeschichte* (so far 2 vols.); E. Stromer v. Reichenbach: *Lehrbuch der Paläontologie* (so far 1 vol.); F. Broili in the new edition of the *Grundzüge der Paläontologie* of Zittel (so far 1 vol.). The great work of Zittel and Schenk is likewise entirely based on these methods.

has been already said applies also for such times when the organisms in question had the best opportunity to demonstrate their evolutionary capacity—for instance, the Pteridophytes in the Carboniferous age. That the life conditions at that time were the most favourable is shown by the great number of individuals and species which lived together and successively in those days.

3. Curious fossil forms like the Graptolites, the Trilobites, Stegocephali, the great Lizards of the Mesozoic period, among the animals, and the Cordaites, Bennettites, and Pteridosperms among the plants, are to be regarded as really extinct forms. Although these may, with some strain, be brought in between two existing classes, they do not become absolute 'links' in the sense of ancestral intermediates. They appear and disappear as Trilobites, Stegocephali, Bennettites, etc.

How entirely unjustifiable it is to see 'transitional forms' in all forms which cannot be identified with any recent class or order is shown by the few remains of some formerly widely distributed organisms. In the Mesozoic age the Ginkgo trees form for the time the predominant Gymnosperm group; they are a well-circumscribed peculiar group on account of the singularity of their leaves—which cannot be mistaken for the leaflets of our Conifers nor for the gigantic leaves of the Cycads (the two other main forms of the Gymnosperms)—and on account of the habit of their stems. They have remained to this day since the Permian era as they were. What descendants of these should we have sought for among our Conifers or Phanerogams, if the solitary surviving species, *Ginkgo biloba*, had also become extinct and had been buried in some inaccessible place or had not become known to Europeans as still existing?

What would have arisen from the Rhyncocephali of the Permian system, or from the Nautiloids and Crinoids if they had been more numerously preserved, we can easily presume from Hatteria (*Sphenodon punctata*) and the few Nautili and Crinoids which are still living. They would have remained the orders of Rhyncocephali, etc., as in fact is the case with the only Hatteria which represents the entire value of the order [1] (Rhyncocephali) with its widely branched relatives of the past.

As it occurs to no investigator to regard the present Hatteria as an actual transition from the Newt (which it most resembles) to the Crocodile, just so is no one justified in regarding the Permian Rhyncocephali as phyletic linking forms, as still, however, always happens.[2]

The same remark applies to the Permian Stegocephali, which for the time being are only known as 'Stegocephali'—i.e. as animals which, like the Amphibia, possessed a free living larval form and two occipital

[1] *Hatteria punctata* forms in the textbooks the order of the Rhyncocephali, although to-day it possesses the systematic value of a good species. *Ginkgo biloba* represents in the botanical system an entire class, although all the still existing individuals are so alike that they only form one systematic species.

[2] R. Hertwig says, for instance, in his textbook, p. 590: 'In the same way the "Rhyncocephali" lead also to the Hydrosaurians, particularly to the Crocodiles, since a double cheek-bone exists (*jochbogen*), and the "quadratum" is firmly attached to the skull.'

bosses, but, with these, reptilian-like teeth and a scaly covering, since of an actual connection of our recent reptilian orders with the Permian or later Stegocephali nothing is known. Suppositions might be permissible if the evolution for similarly varied animal groups (as are the present Reptilia and those Stegocephali) were demonstrated or at least were shown to be very probable.

Depéret chastises excellently this 'method of approximate estimation.' It consists therein that for a genus of living or recent animals, whose genealogy it is desired to ascertain, several other genera are selected from the series of earlier geological periods which present a certain analogy to the first in the structure of an organ or in a small number of organs. With the aid of these genera a series is arranged which, with regard to the changes of the organs taken into consideration, appears to fit in with natural evolution. For the Mammalia, for instance, there is taken as the touchstone, sometimes the structure of the molars, sometimes the canines, here the progressive regression of the side toes, and there the graduated development of the nasal bones, the horns, or antlers, while the rest of the organization is almost entirely neglected. 'Furthermore, the chronological order of the appearance of the fossil forms, which are serially arranged, does not cause over-much embarrassment in these cases.' Thus have Gaudry and Boule, in constructing the pedigree of the Urcidæ, between the Hyænarctos out of the upper Miocene and the first Bears (Ursus) out of the Pliocene, placed a still living genus—the Aeluropus (from China)—because this animal, with regard to the advanced development of its molars, represents a transitional grade between the two genera which it is endeavoured to connect together. Such anachronisms are to our mind quite inadmissible.

The same anachronism everyone is guilty of who inserts the still living Rhyncocephalia (Hatteria) in the same way, not only between genera, but between two orders (Newts and Crocodiles), as a connecting link.

4. Furthermore, the actually established continuous reduction of the sexual generation in the higher plants, determined by comparative botanists and advancing from the Ferns to the Angiosperms, does not prove by itself alone that Ferns, Gymnosperms, and Angiosperms form a genetic series. This is because also within the Pteridophytes themselves there is determined a similarly increased reduction, if we assume a suitable association—i.e. one which is based upon this character. From the macrospores of the Water Ferns (e.g. *Salvinia*) there is produced a very small female prothallus which no longer leaves the spore integument ; the microspores even only form a pair of cells ;[1] they are therefore, so far as reduction is concerned, not behind the Gymnosperms. Nevertheless, the Water Ferns remain true non-seed-bearing plants, since the macrospores are thrown off before the formation of the embryo and therefore do not leave the mother plant as a so-called seed.

[1] E. Warnung : *Handbuch der systematischen Botanik,* p. 144.

Even within the class of Club-mosses a similar procedure is to be observed.[1]

Heterospory finally—i.e. the peculiarity of forming small and large spores for separated male and female prothalli (which has been regarded as a progressive stage in the evolution of the whole plant world)—is found in the most varied plant groups ; there are homosporous (with spores of only one kind) and heterosporous forms among the Ferns—the Equisetæ and the Club-mosses. The present Equisetæ are even exclusively homosporous (therefore ' lower ') forms ; the Calamariaceæ were heterosporous.

Thereby is it demonstrated by evidence that differences in the spore formation and reduction of the sexual generation are not phenomena which by themselves can be taken as transitional stages towards other ' types ' or as constituting a general higher formation.[2]

5. On whatever principles we may proceed in the systematic classification of the present organisms, so far we always arrive at the result that the plants and animals fall into a few larger groups which exist separately but close to each other.[3] Are these groups also really types, i.e. primary animal and plant forms ?

[1] E. Warnung : *Handbuch der systematischen Botanik*, p. 153.

[2] J. Reinke agrees to this when he says (*Einleitung in die theoretische Biologie*, p. 471) : ' If in this sketch I have put forward the results of comparative examination, without further remarks, as evidence of the genetic connection (between Ferns, Gymnosperms, and Angiosperms) of phylogenetic evolution, the reader will not be in doubt with regard to the manifold hypothetical elements in this partly fanciful description.'

[3] With regard to the various botanical divisions see the very instructive chapter on the *Stämme des Pflanzenreiches* by Kerner v. Marilaun (*Pflanzenleben*, II, p. 488. This chapter is not in the second edition.

At present we can only say so much: viz. that those plant and animal groups (families) which not only now but also during the whole of their ancient existence have remained alike in definite characters peculiar to themselves—these we must regard as types.

In this connection it is immaterial what rank they are given in the present classification, whether they form species, genus, family, order, class, or stock. *Ginkgo biloba* is to-day a species, since all the individuals are entirely alike; but according to the above criteria it is a type. That is recognized in systematic classification itself, since it ranks *Ginkgo biloba* as a 'class' in consideration of its peculiarities in comparison with other Gymnosperms.

Hatteria forms a genus with several species, which, however, differ but little; its peculiarities, however, raise its position to that of an order, so soon as it is compared with other (in a wide sense) similar animals (Reptilia), and so on.

It is the object of palæontology and biology to determine the number of fundamental forms of animals and plants. The task is difficult, since we know how much, for instance, parasitism, adaptation to other habitats, and transition to sessile modes of existence, can influence, alter, increase, or decrease the habit and the degree of differentiation.

One thing certainly already appears now to be as good as certain, viz. that at least some fundamental forms in the animal and plant world are firmly retained; for this reason an attribution of all animals and all plants to one fundamental form is 'extremely unlikely' to be correct, as all investigators, so long as they are 'exact,' will willingly allow.

It would be well if in every evolutional hypothesis these suggested limitations were adhered to. There still remains a wide field for research, and especially the question whether and how the types were established within their limits, and what they were before they appeared as completed types which would be preserved for us and perhaps could alone be preserved. Should, for instance, the Ferns represent a true type, yet that is not to say *a priori* that the plants which we call Ferns were always so constituted. One thing alone seems fairly certain. 'Ferns' are and were always different from Equisetums, or 'Invertebrates' show no genetic connection with Vertebrates, or Vertebrates were never such Invertebrates as we know them.[1]

[1] Reinke agrees with this quite emphatically. In his book, *Die Welt als Tat*, p. 351, he says: 'It is of the greatest significance that in the multiformity of forms almost unlimited types appear. These types embrace the enormous number of the now living plants and animals and those which have reached us as fossils.' On p. 352 he continues: 'I quote here a palæontological fact, which is of the greatest importance for the theory of descent. While we find in the petrifactions of the Palæozoic and Mesozoic periods not only other species, but also a preponderance of other genera than at present, yet according to the evidence of the remains found there has never been discovered, even in the oldest periods, any other main type of animal or plant than what we have in the present age. Genera and species have become extinct and been replaced by others. Yet the fundamental types have survived from the time of the oldest formations to the present day.' The acceptance of a polyphyletic evolution—i.e. of varied developed series, separated from the commencement—becomes more and more the dominant opinion. O. Hertwig, Reinke, Kerner v. Marilaun, Steinmann, Zittel, Depéret, Koken, Wasmann, and others regard this as the only admissible view or at least as the most probably correct one.

Appendix: Are there 'higher' and 'lower' types?— In the *Biologischen Zentralblatt* (Nos. 1 and 2, 1911) B. Franz published a long article under the title 'Was ist ein höherer Organismus?' Franz concedes in the first place that the opinion, that there are various high grades of development of animals and plants, and that 'man represents the highest grade of organic evolution,' has been held essentially unchanged throughout the whole period of biological research since the time of Aristotle (p. 1). Then, however, he says (p. 2) that all our ideas as regards high and low in the organic kingdom are perfectly objectless. This so far had certainly been recognized by only three biologists (p. 2).

On page 3 it is stated 'that the assumed higher organism is neither more perfect than the assumed lower, nor in principle is it distinguished from it by additions or differentiations as of higher grade than the lower . . . wherefore it is best henceforth to avoid entirely the misleading expressions "higher" or "lower" organization, "more perfect" or "less perfect," etc., in biological parlance.' The author will certainly not succeed in establishing this idea, and the number of his disciples who, in the vanquishing of false prejudices, 'go far enough' (p. 2), will certainly not far exceed the three aforesaid.

Franz arrives at his—as he himself feels—extraordinary conclusion because he confuses purposeful (*Zweckmässig*) and perfect (*Volkommen*). We will therefore at least explain these two expressions.

What Franz understands by 'more perfect' (p. 5) he indicates by taking as synonymous the following expressions: 'more favourable; better; more purposeful; adapted in a higher degree; more capable of life than the lower.' These expressions are not at all equivalent.

A thing—for instance, an instrument—we call purposeful, useful (*Zweckmässig* = adapted to its purpose), if it fulfils the objects it should attain to—viz. its purpose. The old church clock in a poor village is useful if it only strikes the hours and half-hours, since it does what it can and must. There is full agreement between its construction and its service so far as the maker has involved this in its works. A modern chronometer which in a whole year does not vary one minute from astronomical time, which shows the days of the week and month, regulates itself automatically against changes of temperature or humidity, is, considered by itself, no more purposeful, since it only furnishes what is instilled into its mechanism; even in this case, it is governed only by simple agreement between construction and service.

Nevertheless, such an exact chronometer is termed by everybody a more perfect instrument than a mere village church clock, since it is more perfect, sufficing in a more purposeful way for several services instead of fewer; it is more perfect, being able to indicate exact time under changing circumstances instead of unchanging ones; it is more perfect since it regulates itself instead of being regulated by other means, etc.

Thus 'perfection' implies suitability to purpose (*Zweckmässigkeit*); its higher or lower grade is determined by the extent of the services which are needed to be rendered simultaneously and purposefully, furthermore by the exactness and rapidity of the fulfilment of the inner capacities.

If what has been said be applied to organisms, many difficulties which Franz presents disappear (see, for instance, p. 9 *re* land vertebrates and fish).

Those animals, therefore, which come into purposeful relations with more external objects, and have more means of making these objects useful, are more perfect animals. Man is therefore the most perfect being.

There is no form of organism which he cannot utilize: he tames the beasts and cultivates the plants. No inorganic energy escapes his service if he needs it. There is no faculty of thought which he has not increased by suitable instruments in order to come into purposeful connection with more objects which otherwise would elude his observation. No medium has remained inaccessible to him: he traverses the water and would make the air his own. Thereto will Franz reply that these be intangible 'values' (*Werthe*). Let it be called what one will, but it is nevertheless an actual reality. By instruments the extent of the purposeful relations of mankind with the external world are in fact increased: man, therefore, becomes ever more perfect. These relations are, furthermore, to a large extent necessary so that he may preserve his life, which is also something real.

No animal and no plant exhibits such a multifarious purposeful intercourse with surrounding nature: they are all confined within very narrow limits to definite objects and definite conditions of life; they are specialized in their entire construction, i.e. adapted in one direction, which is the direct opposite of 'perfect' (*volkommen*).

It is therefore incomprehensible when Franz (p. 36) writes: 'It appears to me, for instance, that by the formation of the intestinal canal, the formation of the foot and the arming of the head, the Ruminants have decidedly assumed a similar supreme position (*Gipfelstellung*),' as has man, in the general opinion, by virtue of his brain. The quadruple divided stomach of the Cow and the horns on the head are certainly very purposeful instruments for a strong grass-eater and an otherwise quite unarmed beast. If man has no specialized formation of the intestinal canal, what he has permits other nutrition than only green fodder, which under the circumstances is very much to the purpose and is recognized by all as 'perfection.' As substitute for the horns, which can only be used in close combat, he has known how to provide himself with firearms, or traps, as opposed to which even the mighty herds of the Bison have at last had to yield. In the invention and manufacture of such extremely and vitally purposeful things his brain has done him the greatest service. His weapons, it is true, do not grow upon him, as do the horns on an ox, but his understanding has 'grown' instead, and the gun is just as much an existing thing, and fulfils in reality the same service, as horns upon the head.

Certainly of 'consciousness' (i.e. of psychic capacities and a suitable conformation of brain), as criterion for 'higher' and 'lower,' Franz will know absolutely nothing. This he expresses thus: 'All disputation about consciousness in animals and plants has remained so far not only hypothetical but unsatisfactory. . . . Therefore every objection based on consciousness I decline to meet as being outside discussion' (p. 11). Franz in this treats the matter really too easily. Such procedure can never lead to the establishment of the truth. A whole series of investigators have occupied themselves thoroughly and experimentally with the question of consciousness: their reasons are to be tested and thereby must a decision be arrived at! One thing is also quite undisputed, viz. that man in any case stands high above all animals in 'intelligence.' This truth is beyond doubt and therefore can and must be regarded as the criterion of his 'highest' position (*Gipfelstellung*), since the intellect provides and provided mankind with actual definable aids to his existence and to the extension of his purposeful relations to the outer world.

B. Franz has thus shown by the evidence that with all organisms of the present and the past there is a perfect agreement between construction and function, between needs and faculties, in order that they be satisfied, that all organisms are purposefully arranged. On the other hand it will in the future be also maintained without hesitation that in the realm of the living there exist the most manifold grades of perfection, for it is certainly no 'illusion' (*Tauschung*) if 'in the case of certain organic creatures we *cannot* mentally separate the values which we are accustomed to associate with their existence' (p. 6). 'More perfect' and 'more imperfect' instruments there will always be, even if both kinds of tools are quite well fitted for their temporary uses.

SECTION III.

EVOLUTIONARY HYPOTHESES.

CHAPTER I.

THE PRINCIPAL ATTEMPTS AT EXPLANATION HITHERTO.

MUCH of that which we shall say in the following pages regarding the evolutionary hypotheses already put forward has only an historical value. An opportunity, however, thereby presents itself for learning the nature of the evidence by which it has been attempted to establish the theory of evolution as opposed to that of constancy. The refutation of the theory of the unchangeability of the systematic species, which with Lamarck was hardly much more than a simple denial, constitutes the one permanent result of the best known of all theories of evolution termed Darwinism and Lamarckism.

§ 1. *Lamarckism and neo-Lamarckism.*

(1) *The original doctrine of Lamarck.*

(i) Short description.—Jean Baptiste Chevalier de Lamarck (1744 to 1829) published in the year 1809 a work entitled 'Philosophie Zoologique,' in which, for the first time, the unchangeability of organisms was entirely denied, and the development of the present organic world from inorganic matter by spontaneous generation was affirmed: the animals from gelatine masses, the plants from masses of mucus. The finest fluids penetrate these masses, make them soft (=cellular) and therefore suitable for life. Then, according to a definite plan determined by the great Creator of all things, there followed the creation of ever more and more complicated forms.

How did Lamarck arrive at such conclusions?

In the first place it appeared to him unnatural that the successive organic worlds (creations), so different from each other, should be destroyed by general catastrophes and then again replaced by a new creation in altered forms. It appeared simpler to him to suggest that the separate 'creations' arose genetically from each other. The variability of the organisms—which is certainly the premiss of all evolution—he sought to show, since he demonstrated to us thoroughly by examples how organs *can* alter, though not that they do so in point of fact. He was strengthened in his opinion by the observation of the similarity of the organic groups, which was most easily explained by a common origin. Furthermore, it struck him how the organs of the animals were so perfectly adapted to quite definite needs, to a narrowly limited mode of existence. The idea appeared to him to be closely associated that it was just these needs which must be the cause that the organs are precisely so constituted, often in a quite wonderful and peculiar way. Lamarck then attempts to make it comprehensible how the animals could arrive at this purposeful constitution of their organs, which are adapted so wonderfully to the most varied needs of existence, even to the smallest detail. That is the chief idea, and, in a certain sense, also the greatest service Lamarck rendered: he puts forward a 'theory of organic purposefulness,' not a doctrine of descent, which all problems of the 'history of life' involve.[1]

With regard to the origin of 'organic purposefulness' he writes as follows:

'That, in the first place, any alteration, even inconsiderable, in the circumstances in which each race of animals finds itself, induces an actual change of its requirements. That, in the second place, each alteration in the requirements of the animals renders necessary other faculties in order to satisfy these new requirements and consequently other habits. That thereby each new requirement, since it renders necessary new faculties to meet it, demands from the animal which experiences it either the extended use of an organ of which it had hitherto made less use, whereby such organ is developed and considerably enlarged, or the use of new organs to which the requirements within it imperceptibly give rise through the efforts of its inner perception' (*Gefühl*).[2]

An example may explain the above:

[1] That we really perceive, in the explanation of the purposefulness of the organisms, the chief merit of Lamarck, is shown in the clearest fashion by the work of the Lamarckian disciples. Thus, for instance, B. M. Pauly says (*Darwinismus und Lamarckismus*, Munich, 1905, 46): 'His works [L.] contain a theory of organic purposefulness that . . . at this moment we have not yet risen above.'

[2] See Dr. A. Wagner: *Geschichte des Lamarckismus*, Stuttgart, 1909, p. 32.

'The bird, whose needs attract it to the water in order to seek its food therein, spreads its toes apart when it desires to strike the water and swim upon its surface. The skin which unites the toes at their base acquires, by this unceasingly repeated extension of the toes, the habit of spreading itself out. Therefore in time the broad swimming webs arise which at present connect the toes of Ducks, Geese, etc. These efforts to swim—i.e. to strike the water in order to progress in this liquid and move therein—have also broadened the toes of the Frog, the Sea Tortoise, the Fish Otter, the Beaver, etc.'[1] Lamarck thus attributes the purposefulness of the organisms to their striving towards the purpose concerned! His doctrine is a final (*finalistisch*) one.

Everything that the animals newly acquire in this manner is, according to Lamarck, inherited by the offspring and thus becomes ever more and more fixed.

(ii) Criticism.—(*a*) We accept much of what Lamarck says, but not always for his reasons. If the catastrophic theory be denied, as in itself an improbable idea, then we must also reject the unchangeability of species. In our Introduction we have entered into details regarding this.

It is also correct that the organisms must alter themselves if an adaptation to changed environments generally be effected. (In contradiction to Darwin,

[1] *Geschichte des Lamarckismus*, p. 35. Many other similar examples are given. The most remarkable and most popularly known examples, which Wagner does not mention, are those of the Kangaroo and Giraffe. *Philos. Zool.*, chap. vii.: *Influence des circonstances sur les actions des animaux.*

according to whom it is the individuals which *by chance* are better fitted which survive.)

(*b*) On the other hand the acceptance of spontaneous generation, independently of the philosophical impossibility, is a serious methodological error. Everything that we observe — and every hypothesis must be based on that—speaks against spontaneous generation (see above, p. 96).

(*c*) The idea which Lamarck has formed regarding the process of the new formation of separate organs, cannot *a priori* be disproved. Lamarck, however, at the most explains how many birds acquire swimming webs, long necks, climbing claws, etc.; he does not, however, explain at all how these animals arrived at the general organic type of 'Birds': since before they acquired swimming webs, etc.—i.e. a part of the entire organism adapted to definite services—they were already birds. The same applies to the other 'types' which we term families and classes. They are now sharply separated from each other and, according to palæontological evidence, were always so; they must therefore be regarded as a 'given' something—not as something which has 'happened' (*geworden*).

Certainly Lamarck, at least according to his words, regards the whole development as due to a plan of the Creator. Therefore we must assume that either from the beginning or from a very early period the said types were established in the primary forms, so that every further development should occur within the limits determined. Thus can Lamarck explain the origin and preservation of the types.

Some modes in which, according to palæontology, the formation of differently constructed organic forms proceeded—viz. increase of size, specialization, regression—can certainly be partially explained as Lamarck proposes.

(2) *Neo-Lamarckism.*

(i) Statement.—We have seen how Lamarck explained organic adaptation. He ascribed to the organism itself the faculty, in the first place, of recognizing in some way the newly arising requirements, of perceiving such, and then, by willed and conscious (?) efforts, of meeting the new needs and altering the organs concerned in a purposeful manner. The adaptation to a purpose which we now see completed before us is thus the result of striving towards the purpose by the organism itself: it is a self-adaptation.

This self-adaptation Darwin has denied, and in its place put the survival of such as, *by chance,* are the fittest. That was a mighty retrograde step as contrasted with Lamarck, as is gradually more and more recognized. The natural historians and philosophers who in the last few years have again, in large numbers, reverted to Lamarck's ideas, term themselves neo-Lamarckians and partly as of the 'psychobiological school.'[1]

[1] As their chief representatives, who also have published formal programmes, we may mention: A. Pauly (*Darwinismus und Lamarckismus*), A. Wagner (*Der neue Kurs in der Biologie*), the *Geschichte des Lamarckismus,* and some smaller writings. R. H. Francé (*Das Leben der Pflanze.*, Stuttgart, 1905–1908) cannot be placed on the same level as the other investigators named, despite Professor Wagner's defence.

What was good in Lamarck's doctrine was taken over by the neo-Lamarckians and defended as victorious over materialism in general and Darwinism in particular —viz. the doctrine of 'Autoteleology' of the organisms, as it is now called.

If, however, neo-Lamarckism be regarded as a theory of evolution, it is a terrible mixture of assumptions and postulates without any comprehensible basis at all. Nowhere, whether in the works of Francé, Pauly, or Wagner, do we find any thorough presentation of the results of palæontological research. In its place they put forward as their chief argument the phrase, that evolution must embrace indiscriminately everything—man, animals, and plants—all of which have been evolved from common ancestors.

This they demonstrate thus :

The similarity of the organisms, especially with regard to the 'psyche,' which alone renders possible and guides the evolution, must be explained by a common origin.

Now, however, all living beings, and perhaps also the so-called inorganic bodies, possess a 'psyche,' and one provided with faculties of recognition, effort, and decision. These faculties are naturally not of the same perfection in plants, animals, and man, but are in no way essentially different from each other. The 'psyche' of plants, animals, and man, presents therefore a single series in which the fundamental peculiarities of all organisms—(viz. decision and will)—are gradationally perfected. They are therefore similar, especially in the 'psyche,' the basal factor of all evolution.

Since, now, each similarity which consists in the possession of one and the same perfection, although in different degrees, must be explained by a common descent, therefore man, animals, and plants have arisen from each other or from common ancestors.[1]

(ii) Criticism.—(*a*) The main argument of the neo-Lamarckians is the assumption *a priori* that similarity, quite generally and without any limitation, depends on descent. That, however, is false, and, in the sense which is imputed to this assumption by the psycho-biological school, utterly impossible.

Proof.—We have already shown thoroughly that

[1] It has not been easy for us to frame an argument *in forma* from the programme writings of Francé, Pauly, and Wagner. Often we stand simply helpless before such expressions as the following, in which Francé endeavours to make us 'comprehend' the existence of the plant soul (see Wagner, *Gesch. d. Lamarckismus*, p. 202): 'The psychic working principle in plants has so far repeatedly shown itself to be of very restricted powers. It will be well always to emphasize this point, since it is precisely thereby that the objection regarding the unjustified humanizing (*Vermenschlichuung*) of plant actions is in advance struck upon the head.' Nevertheless he proceeds: 'that they [actions of men and plants] are alike in principle, which permits the conclusion that they are of like origin. Therefore the chief characteristic of the plant and cell souls is the narrowness of their judgments,' or in other words, 'the many failures and the manifold stupidities which are found in plant life.'—'Plants can be easily deceived, and movements may be caused the futility of which is easily seen by our thought but not by that of the plant.'—'Our brain cells see through the stupidity of our body cells because their powers of judgment stand higher. . . . They have learned to do so because from the beginning they have never done anything else than to practise themselves in judging and thinking, while those (the body cells), as the common working horde, also had to devote themselves to varied handiwork. This affords us the particular key to the spiritual constitution of the plant. [?] The poor thing can only think with body cells; it has no special thinking organ and has therefore been sent to the bottom of the class in the school of life.'

only specific similarity is determined by common descent, as actual practical observation shows. Palæontology certainly renders it probable that also fairly different animals may descend from common ancestors. This difference was acquired through differentiation and specialization, but always within narrow limits. Never, however, do the fossils found demand the assumption that a higher class arose from a lower one, to say nothing of one family arising from another, or, in the extreme, animals from plants.

(*b*) That the 'psyches' of man, the animals and plants, are only a perfection of the same fundamental faculties of all organisms is altogether false. In the first place the soul of man and also those of animals and plants cannot be regarded as 'perfections' of matter: they are substantial components of the organisms. There is an essential difference between a 'psyche' which thinks and shows a free self-determination (the human soul) and one whose faculties do not extend beyond the provision of sensitive recognition and sensitive endeavour (the animal soul). With us the senses do not suffice for thought and free-will, nor do they with animals, and never, really never, is there to be observed anything of the sort with them. Has not the entire modern animal psychology been written also for the psychobiologists? (Wundt, Thorndike, Hobhouse, Morgan, Stumpf, Wasmann).

The same applies to the difference between animal and plant souls. That plants respond to the same external stimuli otherwise than do inorganic bodies, that they can adapt themselves thereto, etc., shows that they are something essentially higher than mere matter and even than a machine. That they have power of recognition and conscious power of effort,[1] is, however, contrary to experience, which by all criteria shows that the plants do not sensibly recognize, feel, and will (see above, p. 108).

In the animals, however, together with the self-

[1] This Haberlandt, upon whom we quite particularly depend, has also not shown. What Haberlandt, by his classical studies, has contributed thereto is stated by Wagner (*Geschichte des Lamarckismus*, p. 145), viz., 'direct capacity of adaptation of the plants, power of self-construction, correlative influence in the formation of tissues, the control of the entire plant body in its finer and coarser construction through the function.' All this was exactly taught by the Christian philosophers, often in quite the same words. They therefore ascribe a soul to the plants, but certainly not one acting with consciousness, since all expressions thereof are lacking.

That we are not forced, therefore, to ascribe to the plants (and animals) 'judgment,' 'thought,' etc., no one other than Haberlandt himself has clearly shown. He protests, namely, quite recently, against the exploitation of his words and experimental results by Pauly, Francé, and Ad. Wagner (he gives the three names himself) in the following significant words: 'If the results of the newer stimulus physiology and sense physiology in relation to plants are brought in in the most comprehensive fashion for the foundation of a psychobiology and plant psychology on Lamarckian principles, then this implies an advance in thought which is not justified. The possibility of psychical phenomena in the whole animal and vegetable kingdom can be calmly conceded, without in the very least degree imagining that the most varied self-regulations of the organism, and physiological and morphological processes of adaptation analogous to human efforts towards a recognized goal, can be explained teleologically in the strict sense of the word.' (G. Haberlandt: *Physiologische Pflanzenanatomie*, Leipzig, 1909, 569 A.I.)

That is excellently put. But Haberlandt might have added that his investigations do not justify the ascription of consciousness to plants generally, nor any sensibility, since his 'sense organs' are organs for the reception of special 'stimuli' and nothing else. (See his *Sinnesorgane im Pflanzenreich*, Leipzig, 1908, and *Physiolog. Pflanzenanatomie*, p. 520.) (See above, p. 112.)

regulations, adaptations, etc., which we ascribe to the 'psyche' of plants as their last cause, we note still other faculties which remain entirely inexplicable unless there be ascribed a power of recognition and of spontaneous effort. This recognition and striving power, as observation teaches us, is equal or similar to our own if we act as thinking beings, but incomparable with our intellectual power and that of free-will.

There does not, therefore, exist that continuity of the 'psyches' of plants, animals, and man. Thereby fails the main argument—nay, the only one—in so far as neo-Lamarckism would be a general, all-embracing, and explanatory hypothesis of evolution.[1]

(*c*) Neo-Lamarckism may explain the adaptive faculties but not the perfection of organization of the various organic types: these are something 'given,' and remain so during the whole geological period of evolution.

Webbed feet, long necks and long tongues, climbing claws, etc., may be explained by adaptation through purposeful efforts, but some sort of neck must have been possessed already by the animal; and to some one type must Bird, Mammal, and Serpent have previously belonged.

[1] Francé (*Pflanzenpsychologie als Arbeitshypothese*, p. 23) confesses that neo-Lamarckism (as he, Pauly, Wagner, and others represent it) stands and falls with the continuity argument. 'Our working hypothesis rests before all on the argument of continuity. It stands or falls with the doctrine of Evolution. [(?) This should be: 'stands . . . with this argument'; since with the doctrine of evolution as a fact would fall, it might be assumed, every evolutional hypothesis (as an explanation).] The plant descends from the same original being from which man also has been evolved.'

How did the organisms arrive at these constructive plans?

That appears to Wagner himself, who otherwise, once at least, ridiculed 'types' and 'constructive plans,' to be a difficult problem: 'It appears to be a very difficult problem, perhaps the most difficult in the whole evolutional doctrine'[1]—and, as we can add, a problem decided for the time being by palæontology against Wagner, since the organic types do not develop the one from the other—they are simply *there*.

Despite the evident methodological errors—it is, as an evolutional hypothesis,[2] a purely theoretical construction *a priori*—and despite the actual errors (*re* the psyche), we have discussed neo-Lamarckism more in detail because it appears to be destined in the next decades to become the credo of the neonistic and atheistic evolutional theorists. All too long it will not triumph, since its fundamental opinions, to which, according to Wagner and Francé at least, atheism[3] contributes, will never be the common property of mankind.

[1] A. Wagner: *Geschichte des Lamarckismus*, p. 231.

[2] Wagner's book (and also that of Pauly and Francé) is really a defence of vitalism against materialism—e.g. Darwinism—and in this connection he does excellent work in some parts of the book. 'Evolution' only plays a rôle, in so far as 'Pan-psychism' is the principle from which the history of evolution can be deduced. Wagner says this distinctly (*Geschichte d. Lamarck.*, p. 127).

[3] In an altogether hostile fashion Wagner polemically attacks (see above work) the belief in God and divine intervention in evolution; all such things are to him 'belief in miracles, mysticism, and metaphysics.'

§ 2. *Darwinism and neo-Darwinism.*

(1) *Darwinism.*

(i) Darwin's Doctrine.—Charles Robert Darwin, as scientific associate, accompanied an expedition to America on board the *Beagle* in the year 1831. Two observations which he made there gave him particular food for thought.[1]

In the plains of La Plata and Patagonia he discovered fossil remains of gigantic Sloths (Edentata=toothless), especially of *Dasypus gigas*. Might not the small living forms of the Sloth, which are now exclusively found in South America, be the offspring of those gigantic forms? In that case they would certainly have been considerably altered.

The farther he went from North to South, so much the more it struck him how, particularly the Birds (he was an ornithological expert), but also other animals, gradually assumed a somewhat different appearance. Might not this difference be a simple result of somewhat changed environments? In that case the organisms were again variable, and again this was due to the influence of external environmental conditions.

Full three and thirty years he employed after his return (1836) in experimenting—that is, by breeding—to demonstrate the variability of organisms and discover the principle by which this variability was governed in

[1] See as regards the following remarks the excellent report of R. de Sinéty, S.J.: *Un demi-siècle de Darwinisme* (*Revue des Questions scientifiques*, 1910).

nature in order to produce ever better adapted and higher forms.

Since in nature man does not, as in breeding experiments, supervise the selection and thereby fix a desired change, he consequently sought another selective factor, and found it in the so-called 'natural selection.' He published his opinions in the well-known book 'Origin of Species by Means of Natural Selection: or, the Preservation of Formed Races in the Struggle for Life' (1859), and 'The Descent of Man' (1871).

The progeny of the same parents are, according to him, never perfectly like the parents or each other: there are always differences, favourable or unfavourable, among the individuals concerned. All the progeny cannot survive: if we think of the thousands of seeds which a single plant can produce, only those which, *by chance,* are more favourably constituted have a prospect of preservation; they alone succeed also in reproduction. The favourable variations, furthermore, are transmitted. Through some of the offspring of these already better adapted individuals there occurs, again *absolutely by chance,* an increase of the favouring peculiarities; and so it continues.

It is seen (*a*) that the first original and each separate increase of a favourable change as regards struggle for existence happens by pure chance; there occur also at least as many unfavourable ones; (*b*) the favoured individuals become preserved because they alone survive and reproduce themselves; (*c*) the characters which have once shown themselves as favourable are inherited and need therefore only to be enhanced.

It therefore appears as if the living organisms adapted themselves purposefully by their own initiative. That, according to Darwin, is a deception.

Of adaptation striven for nothing is said. Among the individuals varying without object or plan there must however be, so he says, still some at least which by chance indicate an improvement. The entire apparent selection only requires that not all variations shall be favourable, but only some, which are then preserved.

Of a plan in the evolution of organisms, of a Creator who in some manner established this plan in the organisms, nothing can consequently be said.

(ii) Criticism.—[1] (*a*) Darwinism, regarded as a general hypothesis of evolution, explains absolutely nothing regarding what the organisms actually possess, but only why they have not certain characters. 'To maintain that certain organic qualities can be explained by natural selection is indeed, to use the words of Naegeli,

[1] We treat here of Darwinism as it has finally shaped itself. Darwin himself was clever enough not to express all consequences. He also did not exclude the influence of environment. Nowhere did he express himself clearly regarding the origin of life, and even the name of the Creator appears in his works. But all, even the most absurd consequences, lie established in the system, and the most impossible of all—namely, the descent of man from a primitive primary form, by natural selection alone—he has finally (1871) himself deduced and had the sad courage to publicly advocate. Darwin thereby has united his fate with that of his theory and with it become bankrupt. Darwin's doctrine has often been criticized, beginning with the noble Wigand, K. E. v. Baer, E. v. Hartmann, up to our own time. See Wagner, *Geschichte des Lamarckismus*, chap. iii. ; H. Driesch, *Philosophie des Organischen*, I, Leipzig, 1909, p. 260.

precisely as if one, to the question, "Why has this tree these leaves?" should reply, "Because the gardener has not cut them off." That would naturally explain why the tree did not possess more leaves than are actually on it, but it would never explain the presence and origin of the existing foliage, nor do we understand in the least why the bears in the polar regions are white if we are told that bears otherwise coloured could not survive.'[1]

(*b*) Each organism is an harmonious whole, in which the most varied parts are combined into a true unit. It is unimaginable that a symmetrically and harmoniously built complex organism, in which one part is incapable of existing without the others, could build itself unless in the hypothetical commencing form all later organs and parts commenced to form simultaneously, and simultaneously and always perfected themselves in the most complete harmony with each other. That, however, Darwinism will not and cannot concede.

If it be also considered that this developing organism at each stage—not only now—must have been perfectly adapted to the conditions, because otherwise it could not have lived, it is clearly seen that it can only be a matter of accidental variations of some already perfectly developed organs.

Thus, in brief, without a planned total development no organism could construct itself from simpler forms; all organisms at all times were just as exactly suited to

[1] H. Driesch: *Philos. d. Organischen*, I, p. 263.

their environments as those of to-day, otherwise they could not have lived ; the changes in them which have been observed are transformations *within* their type, which type, as palæontological and present observations show, they thoroughly retain.

(*c*) Also these accidental changes, which do not go beyond the particular organic type, cannot be explained by natural selection. If, for instance, we accept that some individual snakes become venomous ones, the formation of a poison apparatus is necessary—a hollow or grooved lengthened tooth, a poison gland, and a connection between the poison gland and the tooth.

This apparatus would obviously only become really of service to the venomous individuals concerned, when for the first time it acted as such functionally; previously, however, it would rather be detrimental. How came it then—on the assumption that the poison apparatus was really gradually developed—that tooth, gland, and canal became harmoniously formed without actual utility ? That evidently happened without the aid of natural selection—rather, indeed, against it : that was effected by the organism itself by virtue of an innate principle of purposeful striving.

Natural selection, therefore, explains no single positive acquisition which is in any way complexly formed.[1]

(*d*) *The 'survival of the fittest,' which Darwin accepts as the principle of higher development, is not proved.*

[1] Compare the example which Driesch (*Philos. d. Organischen*, I, p. 269) carried out thoroughly on Darwinian lines, viz. the acquisition of the regenerative faculty in Lizards.

If a whale, for instance, opens its enormous jaws and swallows thousands of small crabs, small fish, algæ, etc., are then the 'fittest' in any way spared? If a pig seeks acorns, are the 'fittest' spared? etc., etc.

(*e*) *Darwinism has not withstood the proof of palæontology.*

The classes and families are at least, as wholes, found thereby to be precisely so separated and, in conjunction, forming systems as they do to-day. The mixture of adapted and less adapted forms has not been found to exist.[1] The changes by convergence and specialization cannot be comprehended by the mere accumulation of the most minute changes without plan and guidance.[2]

[The words 'specialization'—i.e. change in a determinate direction—and 'convergence' likewise imply this.]

We refrain from further refutation—owing to the impossibility that definite characters, arising by pure chance, could long maintain themselves against the free crossing in nature—of the denial of an intervention of the Creator both in connection with the origin of

[1] On the Darwinian basis it is entirely incomprehensible that such a thing as a system of organisms, constructed of complex and less complex constituents, can exist! According to the Darwinian doctrine there could only be amœbæ: and yet the system is not yet chaotic—as it must be at least if the *chance* theory be accepted—but is really a 'system' (H. Driesch, *Philosophie des Organischen*, I, p. 268).

[2] Depéret-Wegner: *Die Umbildung der Tierwelt*, p. 38. 'Do I go too far if I conclude therefrom [i.e. from the discussion on Darwinism] that, at least palæontologically, the question of the origin of species in their full extension remains unsolved?' See also Steinmann: *Die geologischen Grundlagen der Abstammungslehre*, Preface.

life and the plan of evolution and determination of its limits, etc.

(2) *Neo-Darwinism.*

(i) Doctrine.—As chief advocate in this direction ranks Prof. Weismann (Freiburg i/B.). According to him, natural selection acts not on the complete, fully-grown organisms themselves, but on the 'determinants'—i.e. on the hypothetically smallest material parts in the nuclear cells. Each quality of the organism has, according to him, a determining part in the sexual cells. The determining parts (= determinants) show that variability and that divergence which Darwin imputed to the organism itself. Those best nourished prevail over the weaker. They evince, however, that influence in the grown organism only if the change in the determinants concerned has reached a sufficient grade, so that they then, in the organism, immediately produce a perfected new organ as the first expression of their influence, since it is only when the organism has the new organ in a complete form that natural selection sets in.

(ii) Criticism.—The determinants are hypothetical creations; if, however, they exist they must also develop according to plan and harmoniously and always in the same direction, in order (for example) to found a poison apparatus, just as must be the case in a grown organism.

By 'Weismannism,' therefore, nothing is gained, and the whole question is referred to the microscopic stage, from which we learn no more than from the matured organism.

CHAPTER II.

SUGGESTIONS FOR RELIABLE HYPOTHESES OF EVOLUTION.

THE main point which prevented investigators, like Cuvier and his disciples, from accepting a genetic connection between the present and previous animal and plant forms of different appearance, was the conviction that the organisms always retained their specific peculiarities. It could not be conceived that the earlier living beings had so greatly changed that they could often only be allocated to the same family or even sub-order, in a few cases even to the same genus, but only quite exceptionally to the same species as the present ones.

The constancy of the systematic species of the present is, however, not established; rather are there facts which directly indicate a capacity of variation or can be satisfactorily explained only by acceptance of such.

§ 1. *Direct observation and the facts of animal and plant geography.*

(1) *Direct observation and experiment.*

Breeding experiments show, in the first place, that under purposeful supervision of the reproduction very varied forms can arise, which often remain constant—as, for instance, dogs, cattle, pigeons, and the remarkable forms of the cabbage. As in ‘evolution,’ these alterations can certainly hardly be directly considered, but they show that the organism is no stubborn unchangeable form.

Observations were made and experiments confirmed that the external conditions of existence—such as climate, particularly the temperature, and nutrition—could induce such transformations and fix them permanently in the organism, as would suffice to form various ‘species,’ even if the origin of the changed forms had not been known.[1]

Thus there resulted from seeds of one and the same mother plant, which were sown partly at high elevations and partly at lower and partly in the valley, plants of fundamentally different external appearances. The alpine forms were more congested, hairier—as protection from cold—the leaves smaller and darker green, the flowers fewer and more intensely coloured, than those of the valley. Were the alpine forms subjected for a long period to the same conditions, these changes could become permanent, but the tendency is immediately shown to retain the new characters even if their seed be sown again in the valley.[2]

In Angora not only the goats, but also cats and dogs have fleecy wool. Pure-bred dogs cannot be

[1] See, particularly as regards the following remarks, R. Heffe: *Abstammungslehre und Darwinismus*, Leipzig, 1904, p. 96.

[2] The author could convince himself of this in Innsbrück, where, in the Botanical Garden, mountain and valley forms were shown as progeny of one and the same mother plant (Brassica). From their appearance no one would have presumed so close a relationship.

maintained continuously in India: they become smaller, slenderer and with more pointed features; an exception is the spaniel, which can be kept thoroughbred for a long time by breeding. 'The Hares,' says Ch. Depéret,[1] 'in the level regions of North and Central France always differ from those of Provence in the South by being of larger dimensions, having longer and thicker hair, long and thickly haired ears, and a darker fur. These differences become more emphasized as we proceed from Provence towards Africa. The Algerian Hares are scarcely more than half as large as European ones, while when the Sahara region is reached the Hares are very small and of an "Isabelline" colour. . . . Similar specific differences occur in the Foxes and Weasels of the North and South of Europe.'

The composition of the soil is of great importance for the development of plants and more or less for all plants in the same direction, hence we speak of the desert, salt, and limestone flora. The shells of snails are certainly influenced by the composition of the soil: on soils rich in lime they are thick and wrinkled, on silicious ones thin and smooth.

Ordinary garden cress, frequently watered with salt water, acquires fleshy leaves, like many plants which grow on the seashore.

Instructive are the experiments which have been made with regard to the influence of temperature. Our Woodnettle Moth (*Vanessa levana*), for instance, has two forms, spring and summer. If the chrysalis

[1] *Umbildung der Tierwelt*, p. 116.

of the spring form, which normally hibernates and is therefore exposed to cold, be induced to develop by artificial warmth, the summer form emerges, while the chrysalis of the summer form, exposed to artificial cold, yields the spring form. Thus the difference between the *Vanessa levana* forms, which is rather striking, would presumably originate through the influence of warmth.

Even the method of reproduction, as for example with fungi, can be influenced by change of temperature.

Another example is afforded by the Fireflies: these, both in Germany and the Riviera, show two different varieties which, as experiment has proved, are due to warm influence. Since warmth generally accelerates maturity, so we find, in snails, that the shell shows one or more turns less under warm conditions, while cold and a very favourable environment can defer maturity and thus give the animals time to add under some circumstances one or more spiral turns to the normal number.[1]

Nutrition has also influence on the formation of animals, independently of the directly detrimental want of such. Thus the colour of many birds may be altered by the kind of food given.

The collective terms, indeed, such as midland, or northern flora and fauna, etc., which indicate to the botanist or zoologist a quite definite and objectively

[1] Depéret-Wegner: *Die Umbildung der Tierwelt*, p. 123. That is very important for the judgment of palæontological finds, in which the whole transformation is sufficiently often confined to the enhancement of such deviations.

determined habitat or such peculiarities as only there present themselves, show that the influence of climate, temperature, and soil constitution, affects the construction and nature of organisms, and, as experiment shows, can also alter the form.

Supplementary Note.—Nature of the variations.

Darwin, as is known, accepted variations which were not in any particular direction and of a minimal character, among which natural selection should act. More recently there have been made observations and experiments which show that in many individuals considerable variations can suddenly appear, the constitution of which can be to some extent estimated beforehand.

Thus H. de Vries [1] observed that with the Evening Primrose (*Oenothera Lamarkiana*) within a short period seven new species arose. Such sudden deviations from the parental type, apparently without any external reason and which remain constant, he termed 'mutations.'

Mendel's [2] experiments have shown that by crossing two races of plants differing in one or several characters

[1] *Die Mutationstheorie*, I, Leipzig, 1901, p. 151.

[2] Upon the experiments of Abbé Gr. Mendel (published 1865 and 1869, but then forgotten and again 'discovered' in 1900, simultaneously by Correns, de Vries and Tschermak) there is based an entire literature. Mendel's work has newly appeared in Ostwald's *Klassikern der exacten Wissenschaften*, No. 121, Leipzig, 1901. See Bateson's *The Methods and Scope of Genetics*, Cambridge, 1908; and Mendel's *Principles of Heredity*, Cambridge, 1909; W. Johannsen's *Elemente der exacten Erblichkeitslehre*, Jena, 1909; C. Correns, *Uber Vererbungs Gesetze*, Berlin, 1905; J. P. Lotsy, *Vorlesungen über Descendenztheorien*, I, Leipzig, 1906, p. 99.

(Peas, for instance, with different flowers or colour of seed), new combinations can be formed according to mathematical laws, which, when care is taken to ensure self-fertilization, are constant. The experiments render it probable that accidental qualities, such as size, colour, length of life, and many others, are connected with certain corporeal parts (*Gens*) and maintain an independence in the organism. If, for instance, there be crossed a dwarf race with a very large one, it may happen that all the progeny may be large or all dwarf, according to whether the 'gen' of the 'large' or that of the 'dwarf' becomes utilized.

Mutations and Mendelism both are adverse to the smallest scarcely appreciable deviations. The greatest hopes have been built upon Mendelism, which, however, have mostly been unjustified. Crossing occurs certainly in free nature generally only between individuals of the same species. Most of the systematic species do not admit of fruitful intercrossing. *Lepus timidus* (the Hare) never crosses with the northern *Lepus variabilis*; the two species of Sparrows, house and field sparrow, never cross; horse and ass as a rule only by use of artifice, etc. Just so there has hitherto been no cross between apple and pear, despite their near relationship.

With plants it is a general rule that self-fertilization should be avoided (lowest limit), but, almost equally so, crossing between species (highest limit).[1] Some

[1] 'The union of sexual cells follows, as a rule, only if they are derived from individuals of the same species' (Strasburger: *Lehrbuch*, p. 264). 'Self-fertilization' is mostly only an 'aid in need' (same author).

exceptions, however, are certainly known (Quince and Apple). Hence, by crossing, it is only a question of new combinations of specific characters which are of quite subordinate nature, such as colour and size. In order that the new forms should remain constant strict in-breeding is necessary, which, in most cases, can only be artificially ensured. Consider only Mendel's experiments. That by crossing the 'analytical formula of an organism' may be ever discovered is a dream.

Some palæontological series and many observations of the present organisms permit of recognition that even quite gradual transitions lead from one to another species, especially the evidence of convergence and 'specialization.'[1]

All these observations tell us obviously only how, by suddenly appearing mutations, by crossing and by gradual change in determined directions (caused indeed by gradually changing environment), new forms appear within the limits of definite organic types.

(2) *Suggestions derived from animal and plant geography concerning the origin of transformations.*[2]

It is not all deviations which appear in the progeny of the same animal or plant parents which can be imputed to the influence of changed environment.

[1] Depéret-Wegner: *Die Umbildung der Tierwelt*, p. 138.

[2] See J. P. Lotsy, *Vorlesungen über Descendenztheorien*, II, p. 483; R. Heffe, *Abstammungslehre und Darwinismus*, p. 44; A. Weismann: *Vorträge über Descendenztheorie*, II, p. 235; *Einfluss der Isolierung auf die Artbildung*; L. Plate, *Selectionsprinzip und Probleme der Artbildung*, Leipzig, 1908, p. 396.

There are plants and animals which, precisely in the same habitat, or in the same region, often produce spontaneous deviations—i.e. without a recognizable external cause. Such 'bad' species from the beginning constituted the cross of the systematist who desired to give a name to each form and thus 'pulverized,' as Depéret puts it, the animal species concerned, so that eventually nearly every individual bore a special name. Among the plants there belongs to this category the Evening Primrose (*Œnothera Lamarckiana*), the Blackberry, the Hawkweed, and others; among the animals the Vine Snail (*Helix striata*), and some Mussels (particularly *Unio*), are the most notoriously 'bad' species.

These transformations, however, do not go so far that the connection with a common 'basal form' cannot at once be recognized. More extended and more constant transformations established in organisms arise, however, when an animal or plant form occurs in a definite, well-separated (isolated) region with peculiar environment. Then the so-called local races and local species are formed, which, by the exclusion of free crossing with new blood and through long-continued isolation, assume forms which can then only be classed in the same genus, sometimes indeed only in the same family, as their original ancestors.

We will now treat more in detail those facts from animal and plant geography of the present organisms in connection with which the influence of the strictest separation, and the in-breeding connected therewith which has been effective through long periods, has made itself felt.

The conclusions thereby arrived at form one of the main supports of the transformation hypothesis and the best refutation of the absolute fixity of species which was formerly accepted.

The most favourable conditions are presented in the first place by the small oceanic islands which owe their existence to submarine volcanic outbreaks. We will first give a few examples in detail; subsequently we will collect the points of evidence which speak for transformation as actually established.

1. The Azores and Bermuda Islands (both of volcanic origin).

Azores (1400 Km. from Portugal):

69 Snail species, 32 'indigenous'	the rest
53 Bird species, 1 ,,	European.

Bermuda (about 1400 Km. from America):

Many Snails—one-fourth indigenous, the others American.

2. St. Helena (1800 Km. from Africa, 3000 Km. from S. America):

Land Vertebrates entirely absent.

Land Birds—1 species (*Regenpfeifer*) indigenous.

Land Snails—20 species, all indigenous, without nearer relationship.

Beetles—129 species (128 indigenous); they are divided into 25 genera, all indigenous; two-thirds of them are Weevils whose larvæ live in wood (driftwood).

3. Hawaii (volcanic; 400 Km. distant from land):
   Land Vertebrates—2 Lizards, indigenous; one of these forms a genus in itself.
   Land Birds—16 species, indigenous; they are divided into 10 genera of which 5 again form an indigenous family.
   Snails: Achatinella, 20 species, only occurs here.

Islands of the most recent origin show, as yet, as we might expect, no peculiarities: for instance Krakatoa, near Java (risen in 1883). The 20 plant species (1904) belong to 16 families, just as chance brought them together.

These examples show in the first place that these islands became populated from the nearest continent, since they are, as can be proved, of volcanic origin, and therefore possessed no 'creation' of their own from the beginning. There are, furthermore, found quite predominantly only such animals as can be most easily imported—such as snails, whose eggs or larvæ abound in mud—by Waterfowl, and Beetles (Weevils)—whose larvæ live in wood—by driftwood; in addition naturally birds, which also show the least indigenous forms because they are not confined to the islands and can cross more freely.

For the doctrine of descent from the observations given it may be concluded that animals, which are confined to a narrowly restricted region where somewhat differing vital conditions prevail, deviate from their original specific and generic characters through so-called adaptation to the changed environment. That these deviations are most easily determined in a definite direction on such small islands arises obviously from the necessity of close in-breeding between the few individuals which chance has brought together.

It can furthermore be established that likewise entire continents, which have been isolated since very long periods (Australia), or only in the younger geological periods, were again connected with others (like South and North America in the Tertiary era), possess quite peculiar animal worlds (like Australia and South America proper) and only on the frontier regions show a mixture (for instance, the extreme North of South America and Central America).

For Australia the position is considered to be approximately this : Australia became separated when the most primitive forms of the Mammalia first appeared—i.e. the Cloaca [1] and marsupial-like forms—or, what is much more probable, this continent possessed at the time of its isolation only Marsupials and Cloaca animals and received also no other orders of Mammalia, or again (in the style of the popular doctrine of evolution), no

[1] Cloaca signifies in zoology the common region into which both the urogenital system and the digestive canal open ; all other orders have for both systems separate external openings : 'cloaca' therefore signifies a trifling degree of differentiation which served and serves as the best criterion for the determination of different grades of perfection. The ant-eater (Echidna) and the Ornithorhynchus belong to this group.

higher types were developed from the lower.[1] A fact it is that with the exception of forms introduced by man and of such species as, like mice, bats, and seals, can easily make their way from island to island,[2] no other mammalian orders were found in Australia. What is remarkable, however, in this connection is the circumstance that the Marsupials assume the most varied modes of existence and suitable construction of the body: there are flying marsupials, which resemble the flying squirrel; root and plant-eaters—i.e. the Wombat (similar to the Marmot); insect-eaters with the jaw of the Sorex (? Shrewmouse) or Hedgehog; Carnivora—for instance the marsupial Marten with the typical equipment of the order of Carnivora; springers—e.g. the Kangaroo (like the Steppe mice).

All are Marsupials in other respects, however, constructed precisely like the representatives of the other mammalian order, whose mode of existence they pursue.

The significance of these facts appears close at hand. The present forms, which obviously, as regards everything connected with a quite special mode of existence, are constituted to one end (*einseitig*), do not give the impression of original animals, since these, according

[1] That the Allotheria of the Trias—which are considered as being the first Mammals—are the ancestors of the higher orders, is in the meantime a simple assumption. They are also 'unfortunately very imperfectly known' (R. Hertwig: *Lehrbuch*, p. 590). According to Depéret the remains cannot well be otherwise considered.

[2] R. Hertwig: *Lehrbuch*, p. 591. The Marsupials are confined, except the South American Kangaroo Rats, to Australia; earlier—in the Secondary and Tertiary periods—they existed in Europe and North and South America.

to the evidence of palæontology in each group, are not yet so specially constituted, but show in jaw and other equipment a more general character. It is this circumstance precisely which enables them to nourish themselves in various ways and to live in various regions and habitats. Marked specialization—for instance in the jaw and organs of locomotion or with respect to the habitat—will consequently be best considered as really subsequently acquired 'adaptive' characters.

In Australia—which, as a continent, in contrast with the small marine islets, presented opportunities for the most diverse modes of existence—the Marsupials have, therefore, it is concluded, assumed the most diverse forms to fit them. Since, however, they remained Marsupials, therefore this example shows that they present a real type which only varies but is not abandoned. If therefore—which cannot be exactly proved—the present forms of the Australian Marsupials have really developed themselves by differentiation, then they form at the same time a fine proof that certain basal forms are firmly retained. Time and opportunity for the full development of their entire evolutional capacity the Marsupials have certainly had, but other orders they have not produced. The same thing we must assume also for Europe and America if we will be logical, although the palæontological evidence in favour of a genetic connection of definite mammalian groups with each other is lacking.

On other continents—like North America and Asia, together with Northern Europe, which for a very long time were connected across the Behring Straits—the animal world is very similar, since fresh blood was constantly coming in which by continued crossing rendered an alteration in a definite direction impossible.

The objection might yet be raised that the various continents could certainly have had their own fauna and flora from the beginning. But then it is entirely forgotten that the present continents did not always exist: how otherwise could marine animals be found in the alpine strata? Furthermore it is certain that in many geological epochs the animal and plant worlds in Europe and America were almost entirely the same—for instance, in the coal period; that at times, over the entire northern hemisphere as far as Greenland, a warm climate prevailed—in our regions a tropical one, without marked seasons; and that even in the Ice Age with us the Siberian plants and those of the high Alps still lived in the valleys. There is thus proved by evidence that there has been a continuous changing and constant wandering from place to place: therefore one cannot speak of the countries which at present chance to project above the level of the sea as having their own organisms from the beginning.

§ 2. *Suggestive points in the domain of Parasitism and Symbiosis.*

A very fruitful domain for historical evolutionary hypotheses is presented by the manifold relations of various species of animals and plants to each other or of animals to plants. When individuals of different species enter into close mutual relations, then this is based either on utility on one side only or on mutual support. In the first case we speak of Parasitism, in the second of Symbiosis.

(1) *The adaptive phenomena of the Parasites.*

We will confine ourselves here to the true Parasites, independently of the more or less legitimate kinds of life association which do not imply an actual dependence but rather a companionship.[1]

True Parasites are organisms which exist either upon or in other organisms in order to nourish themselves by the living substance or already prepared nutritive sap of the same.[2] With this peculiar mode of life on or in other living beings (termed 'host animals' or 'host plants') the bodily equipment and the needs of the Parasites are in the most perfect accord. The preponderant majority of the Parasites are adapted physiologically and morphologically to their abnormally deviated mode of existence : physiologically by the fact that they necessarily require for the maintenance of their existence the nutrition to be derived from a definite species of animal or a group of related species ; morphologically in so far as their bodily construction is arranged for the acquisition of precisely this nourishment.[3]

[1] A short description is found in all botanical and zoological text-books. Prof. L. v. Graff has treated of *Das Schmarotzerthum im Tierreich* in his work *Wissenschaft und Bildung*, No. 5, and given an excellent summary of the observations made in that connection. With his interpretation of the facts we cannot, however, always agree.

[2] *Ibid.* p. 8.

[3] *Ibid.* p. 9.

From this agreement between construction and needs on the one side, and the mode of existence on the other, there follows nothing in support of any evolution whatever, since adaptability to purpose—and this agreement is nothing else—is a general phenomenon throughout the whole organic world: all animals and plants possess those organs and capacities (i.e. arrangements in plants) required to fit them exactly for existence in their habitats or resorts, and to enter into or be brought into relations with those objects or organisms which are necessary for the maintenance of life and of reproduction.

The closer observation of the Parasites affords, however, several very important points in support of the assumption that their mode of existence, and consequently their construction and their physiological constitution (as means to an end), are something subsequently acquired (secondary) and not of primary origin; or, expressed in the language of the time, the present form of the Parasites in the adult stage, and many peculiarities of the embryological genesis, present the result of a transforming (evolutionary) process—they are 'adaptations.' If that be really the case then certainly the transition to the parasitical mode of life has been the cause of the production of the most manifold forms of animals and plants, which often enough we can only bring into our 'system' with great difficulty.

'Adaptation' (*Anpassung*) therefore does *not* imply every agreement between construction and function.

The meaning of the word expresses this: to 'self-adapt' (*sich anpassen*), signifies, indeed, the production of an agreement by changes which in the majority of cases commence in the mode of life and react upon the construction. Graff expresses this in the words: 'their deviation from the normal mode of life.'

H. Driesch has expressed himself very tellingly regarding adaptation.[1] 'True' or secondary adaptations, according to him, only occur when 'any species of variation in function occurs which agrees with a variation of any one factor of the medium.' 'We call secondary adaptations all occurrences in the domain of form construction and function which serve to restore the disturbed condition on lines which lie outside the realm of so-called normality.'[2]

The type of a bird, for instance, might justly be termed an adaptation if it could be shown what animal produced its wings, etc., by adoption of an aerial existence. So long as that cannot be done, the bird, with its construction and its modes of life, is a given type and not an evolved (*geworden*) one. Until, however, this, the only correct assumption, be everywhere and logically followed up, in scientific parlance, a long time must yet elapse. In the meantime all that is to the purpose is 'adaptation.'

As bases for the assumption that the parasitical mode of life, and the therewith connected bodily construction of the Parasites are acquired, we may state—

[1] *Philosophie des Organischen*, I, p. 185.
[2] *Ibid.* p. 189.

(*a*) From the outset it is improbable that from the beginning there were organisms which, for the preservation of their individual life, were dependent upon the separate vital functions of others, since they themselves lack the capacity of performing them alone.

The fungi plants, generally speaking, can no longer assimilate, and thus cannot fulfil the task which falls upon the plant kingdom taken as a whole. To the existence of an animal there appertains free locomotion and the formation of sense organs; the animal Parasites, however, often dispense entirely with both, with the exception perhaps of feelers.[1] Multicellular animals have also their special functions—in contrast to the unicellular—transferred to separated tissues (organs); many Parasites—which, in addition, in their total habit belong to the Worms or the Crabs, i.e. to fairly differentiated organisms—often lack even an intestinal system of their own.

(*b*) Had there been existent from the beginning Parasites of the present form and mode of existence there would have resulted, at least in some cases, very singular consequences. We know Parasites which must live not only in one particular species of animal, but often even in one particular organ. The worm *Oxyuris vermicularis* can only live in man—with him and in him must it also have been created. The Trichinæ can only maintain life in cross-fibred muscles. A larval form (*oncosphæra*) of the tapeworm (*tænia cœnurus*) can only develop itself further in the brain and spinal

[1] R. Hertwig: *Lehrbuch der Zoologie*, p. 155.

cord of the sheep. The malaria Parasites (some Plasmodium species) would have had, in that case, from the beginning the positive task of causing fever and death to man,[1] since only to that end are they constructed. Only in the red corpuscles of the blood of man or in the intestine of the mosquitoes (some Anopheles species) can the various reproductive stages be carried through.

Have there been from the commencement beings whose existence is necessarily dependent upon a constant exchange of relations between men and mosquitoes?[2] If the man be not there, then the purely vegetative increase cannot happen and therewith the first condition for further development falls to the ground. If there be no mosquitoes then the sexual generation cannot take place, and the play is at an end. The malaria Parasites are thus physiologically formed for this mode of life—i.e. they have the need of so existing and the necessary equipment to attain their object; therefore the spores introduced into the blood have a pointed form, to enable them to bore into the blood corpuscles.

(*c*) Had the parasitical mode of existence been present from the beginning, particularly to its present extent and in its extreme development, then there

[1] According to investigation so far the Malaria Plasmodes occur only in man and as a phase of Schizogonia (= increase by formation of several individuals from one by fission).—F. Doflein: *Die Protozoen als Parasiten und Krankheitserreger*, Jena, 1901, p. 145.

[2] *Ibid.* p. 130. We have to do with Parasites which, 'in the exchange of hosts between two organisms, live in man and a member of the Fly family'; also, p. 147: 'Another mode of transfer than through the mosquito cannot be according to our present knowledge.'

would have been also from the beginning those shapeless forms which we see often enough in Parasites. On this assumption, however, how could it be explained that there are transitions between shapeless Parasites and normal ones—i.e. free-living representatives of a defined type in which otherwise the separation of type appears to be so clear?

The Platodes, for instance, have their tropical representatives in the free-living Planaria (Turbellaria); they possess an intestine and numerous cilia as organs of locomotion which cover the whole body like a garment. There is also known a large group of similar animals which may also be recognized as Platodes and are called Trematodes. They are thoroughly parasitical and provided with an intestine, but without cilia, bearing suckers and hooks which serve to attach them in or outside their hosts.[1] In many forms the really digesting portion of the intestine disappears entirely; there remain only the œsophagus and the anus. Again, other similar animals are the Tapeworms, all parasites, without intestine or cilia; they are also the most extreme parasites of the whole Worm group. One is inclined to regard the three classes cited as representatives of one and the same animal form which, in the free-living Turbellaria (there are also parasitical ones), exhibit the unchanged original type, since there are also among the Turbellaria the one or the other extreme parasitical kind, which shows similar retrogression to that of the worms ranked with the Tapeworms. That

[1] Graff: *Wissenschaft und Bildung*, p. 25.

there are, however, *true* Turbellaria is shown by their different embryonic development from that of the Tapeworms. There would therefore also have been in the Tapeworms a marked alteration in the genesis of the embryo; the adult forms of a Tapeworm and of such parasitic Turbellaria would by themselves form no reason for establishing different classes.

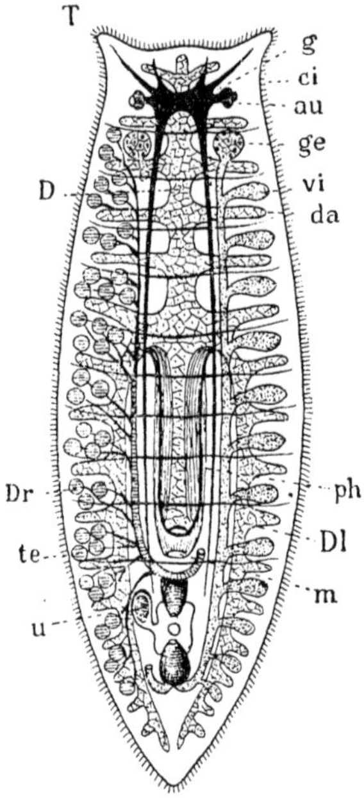

FIG. 32.—PLANARIA.

Plan of a free-living Planaria. au, eye; ci, cilia; D, front main intestine; da, branch of same; Dl, left rear, and Dr, right rear main intestine; g, brain; ge, germinal gland; m, mouth; ph, anus; T, feelers; te, testicle bladder; u, ovarium; vi, yolk gland.

(*After von Graff.*)

As between Tapeworms and Turbellaria so is the separation between Turbellaria and Distoma (sucking worms) not always to be clearly followed. There are free-living Turbellaria with suction apparatus but no hooks, which are only serviceable to parasites. Such Turbellaria could easily become Parasites (Ekto-parasites) if they could thereby obtain nourishment more easily. We have namely supporting evidence by observation which shows how opportunity alone may render an animal a facultative parasite whether the opportunity be artificially or spontaneously provided.

'Thus can animals, which live normally in the excreta of man, also develop themselves inside the human intestine if by chance their eggs find their way there '—as, for instance, the larvæ of the flies *Eristalis tenax* and *Anthomyia canicularis*.[1] The worm *Leptodera appendiculata* lives mostly in rotting material in the soil; if it finds entry into the hiding-place of a snail in close vicinity (which thus shares its habitat) it can also flourish there very well.

Certainly by such examples nothing is shown to demonstrate how a parasite by opportunity only may become a true one which can only exist in a definite foreign organism. We can, in the first place, only say that frequent association at close quarters or under similar life conditions, can often present the most favourable opportunity, and that, with the change of the mode of life, the instinct of the animals and of their offspring, in conjunction with their conformation, may become influenced.

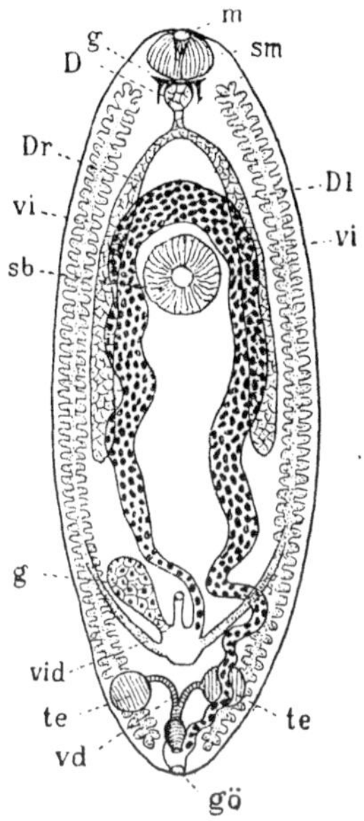

FIG. 33.—DISTOMUM.

Plan of an entoparasitic Distomum. D, anus; Dl, left, and Dr, right main intestine; g, brain; ge, ovary; m, mouth; sb, ventral attachment; sm, mouth attachment; te, testicles; vi, yolk bag. There should be noted the feeble development of the intestinal system (minus outlet) and the strength of the genital one. The sucker is a positive adaptation.
(*After von Graff*.)

The best example how shapeless forms may be associated with normals—i.e. those which agree with a definite type—is that of the

[1] Graff: *Wissenschaft und Bildung*, p. 9.

Copepods (Swimming Crabs). Fig. 34 shows some of the most striking transformations. In the plant kingdom parasitism is also widely extended. Exclusive parasites especially are many fungi and bacteria ; they are often dependent upon 'quite definite animals and plants and even upon narrowly limited parts of them.'[1] Often, however, they can find their nourishment in dead organic remains (saprophytes).

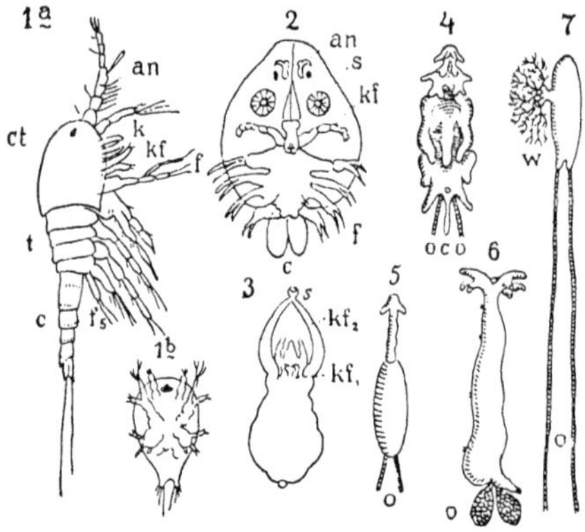

FIG. 34.—COPEPODA.

1. Free-living *Canthocamptus minutus* (*after Claus*): 1a, seen from the side; 1b, nauplius stage. The following are parasitic: 2, Carplouse *Argulus foliaceus* (*after Claus*); 3, *Achtheres percarum*, female (*after Claus*); 4, *Chondracanthus gibbosus* (*after Claus*); *Lernæonema monillaris*; 6. *Lernæocera cyprinacea* (*after Nordmann*); 7, *Peroderma cylindricum*, (*after Richtardi*). Explanation of letters: an, antennæ; c, tail; ct, main thorax; k, jaw; kf, jaw foot; o, egg sac (in 4 and 5, only partly shown); s, sucker; t, breast segments; w, suction roots. The figures show, from left to right, the graduated transformation of the Copepoda type into shapeless bodies. The females show only still the characteristic two egg sacs of the Copepoda. (*After von Graff.*)

With the Parasites in the series of higher plants we see, precisely as with animals, a feeble formation or a perfect absence of those organs which serve for independent nutrition—namely, the leaves and roots. The non-chlorophyllic leaves of the Dodder (*Cuscuta trifolii*) are small yellowish scales. With many tropical parasites the body of the plant is reduced to the flower,

[1] Strasburger: *Lehrbuch*, p. 195.

as with the gigantic flower *Rafflesia Arnoldi* which is seated immediately upon the roots of the host plant. Such extreme Parasites can no longer fulfil the main task of plants—viz. that of assimilation (i.e. to form organic material from the carbonic acid of the air and the water and salts of the soil). By the formation of the most diverse organs, with which they form connections with the host plants, they permeate these and also in most cases destroy them.

A convincing proof cannot certainly be produced that the parasites pursued formerly another and independent mode of life. Yet, also among the plants, we have some indications leading to this assumption. Many so-called semi-parasites like Euphrasia, Thesium, etc., possess green leaves and true roots, but 'they attach themselves by discs or wart-like outgrowths to the roots of their host plants,'[1] from which they draw directly water and nutritive salts. It can well be assumed that the opportunity afforded by coming in contact with other roots when seeking for water gave the impulse to the formation of sucking apparatus. An attraction of the roothairs towards water, whether in the soil itself or in the roots in the soil, must be assumed, since the finest rootlets have always given the impression that they sought for humidity. What, therefore, was formerly caused by chance—namely the formation of sucking apparatus in the said semi-parasites—may gradually become a permanent tendency in the plants, so that thus the Euphrasia, which cannot

[1] Strasburger: *Lehrbuch*, p. 42.

meet with other roots, can now nourish itself but very imperfectly.

(2) *Adaptive phenomena in Symbiotics* (*Ant guests and Termite guests*), *or Myrmecophils and Termitophils.*[1]

'Ant guests and Termite guests' or Myrmecophils and Termitophils are those Arthropods, especially insects and mostly beetles, which regularly live in association with Ants or Termites. Already (in 1894) they numbered 1419 species;[2] at present they may be roundly estimated at 3000, in consequence of the subsequent great advance of this zoological branch of research. We find in these animals adaptations of the most diverse character in the bodily form to the myrmecophil or termitophil mode of life. In the first place we find curious beetles provided with yellow or red bunches of hair on the most diverse parts of their bodies, or with other 'exudatory organs' which are eagerly licked and cared for by the Ants or Termites, and the beetles are often fed from their mouths; these constitute the so-called Symphil type or 'true guest type.' Others of these guests show, in bodily form and colour, often also in the form of the antennæ, a striking resemblance

[1] It appeared to us better, by one thoroughly treated instance, to make the chain of evidence clear to the reader, rather than to give a purview of the whole of the observation material. The study of the Ant and Termite guests belongs to the special domain of P. E. Wasmann, who has occupied himself with the study for decades. The following remarks are entirely from his pen and were written specially for this work.

[2] Wasmann: *Kritisches Verzeichnis der myrmekophilen und termitophilen Arthropoden.*

to their hosts, to whom they present themselves also as their equals; these belong to the Mimicry type which, especially in many of the associated beetles (Staphylinidæ) of the wandering ant of the Old and New World, show an astonishingly high perfection. Other guests, finally, clothe themselves in armour impenetrable by the ant beetles, in order thus to be able to live in their company; these form the offensive (*Trutz*) type of the Ant and Termite guests, etc.[1] The relation of the said facts to the theory of evolution is briefly stated as follows, abstaining from discussing the causes (internal and external) of the evolution and also the manner of it (fluctuating variation, mutation, etc.).

The myrmecophil and termitophil adaptive characters with which we meet in various classes of insects—millipedes, spiders, crustaceans, and in the various organs of these classes, especially, however, in the insect orders of the Beetles and Diptera—convert the forms concerned into proper systematic species, proper genera, and often even into proper sub-families or families which differ widely from their systematic relatives which do not associate with Ants or Termites.

These differences, however, are only to be explained by assuming that in the course of race development by

[1] For more details see Wasmann: *Die Myrmekophilen und Termitophilen*, Leyden, 1895 (*Verh. des dritten Internationalen Zoologencongresses*); *Die moderne Biologie und die Entwicklungstheorie* (1906), chap. x.; *Der Kampf um das Entwicklungsproblem in Berlin* (1907), 1 *Vortrag*; *Beispiele rezenten Artenbildung bei Ameisengästen und Termitengästen* (*Festschrift für Rosenthal Biolog. Zentralblatt*, XXVI (1906), Nos. 17 and 18; *Die progressive Artbildung und die Dinardaformen* (*Natur und Offenbarung* (1909), part 6); *Wesen und Ursprung der Symphilie* (*Biolog. Zentralbl.*, XXX (1910), Nos. 3–5).

adaptation of the said insects, etc., to the myrmecophil or termitophil mode of life, they have developed themselves. Thus do the adaptive phenomena among the Ant guests and Termite guests afford in fact an abundance of evidence in favour of the generic-historical appearance of new species, new genera, new groups of genera, and new families in the animal kingdom.

In support of the above paragraph we will give only a few examples. In the Beetle family of the Staphylids we find in the group of the Lomechusini three genera with altogether twenty-one species, which, by their peculiarly broad bodily form and the arched sides of the thoracic shield, and particularly by the yellow bunches of hairs on the hinder sides of the body, differ strikingly from the rest of the Staphylids (Fig. 35). All these morphological peculiarities are adaptive characters to the true guest relations which connect those beetles with the ants. The myrmecophil adaptive characters form, therefore, the particular reason why these beetles form proper species, proper genera, and a proper group of genera of the Staphylids.

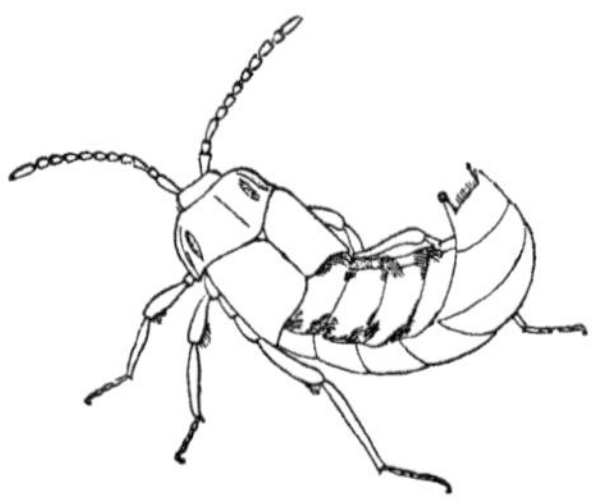

FIG. 35.—LOMECHUSA STRUMOSA F. (mag. 5 diam.).

Furthermore we know so far of over one hundred genera, with about five hundred species, in the family of the Staphylidæ alone, of Ant guests or Termite guests, whose systematic separation also depends upon their myrmecophil or termitophil adaptive characters.

Many of these genera—as, for instance, the Mimeciton (Ant Ape) belonging to the mimicry type of the Dorylinæ guests—are by their adaptive characters so extremely different from their other family relatives, that a systematic sub-family can be justly based upon them, as has been done also for the offensive type (*Trutztypus*) of the genera belonging to the Dorylin guests, *Trilobitideus*, *Xenocephalus*, and *Pygostenus*, which represent the typical genera of the sub-families *Trilobitideini*, *Xenocephalini*, and *Pygostenini*. Very remarkable, too, are the termitophil Physogastric *Aleocharinæ* (Fig. 36), Staphylinids with enormously developed, mostly membraneous posteriors, which can assume the most grotesque forms and most singular positions, as, for instance, in the genera *Spirachtha*, *Termitobia*, and *Termitomimus*. Until now twenty-four genera, with thirty-two species, have been discovered of these interesting creatures in the tropical termite nests where they are eagerly licked by their hosts on account of the exudation of agreeable secretions which here are an element of the blood fluids of the guests, and in return are fed from their hosts' mouths, as is evident from the formation of the tongues of the beetles concerned. The morphological, generic, and specific characters of these hemiptera show themselves thus to be termitophil adaptive characters.

FIG. 36.—TERMITOBIA ENTENDVENIENSIS TRÄG. (mag. 5 diam.)

There is also a particular family or sub-family of small myrmecophil beetles which have been named

Club Beetles (Clavigeridæ) on account of the peculiar form of their antennæ. These differ from their nearest relatives, the 'Feeler Beetles' (Pselaphiden) by a series of adaptive characters fitted for the true guest conditions; by the marked development of the first upper rear segments, which bear a broad and deep exudation groove; by the yellow hair bunches on the hinder parts at the sides, or on the wing cover points; by the shortened, thickened, and very variedly formed antennæ, as also by the retrogression of the feelers which, as organs of independent search for nutrition, have become useless. Classic examples of Club Beetles are our small yellow ones (*Claviger testaceus*) as are the great Madagascar Club Beetles with stag-horn feelers (*Miroclaviger cervicornis*). We know already at present forty-two systematic genera of Clavigeridæ with 124 species, whose generic and family characters are plainly myrmecophil adaptive characters.

There is, furthermore, one particular beetle family—the Paussidæ or Feeler Beetles—which are all Ant guests but nevertheless belong to different biological classes. By far the most of them are true guests which are provided with multiform reddish yellow hair bundles, exudation grooves, and exudation pores which are licked by their hosts. Their very thick and only two-limbed feelers present the most varied and grotesque forms, which, however, all—like the exudatory organs just mentioned—stand in the most intimate connection with their true guest relationship. The species and genera of the Paussidæ and the whole family itself are what they are by virtue of their myrmecophil character.

We know so far 333 living species, divided over sixteen genera. In addition there are five extinct species, of which four lived in the lower Oligocene and therefore in the first third of the Tertiary period. Three of them belong to the still living genera Arthropterus and Paussus; one, *Paussoides*, is only known as a fossil. There is also a Paussus of the Diluvial period (preserved in Copal). The Paussidæ show the nearest systematic relationship with the Carabidæ and with the group of Bombardier beetles (Brachyninæ) to which the most primitive Paussidæ genera closely approach. Since the Carabidæ appeared already in the Trias and in the Lias and since the Brachyninæ are already represented in the upper Chalk and the lower Oligocene, hence they, palæontologically, are to be assumed as the ancestors of the Paussidæ.

As with the Staphylinidæ, Clavigeridæ, and Paussidæ so it is with the Gnostidæ, Ectrephidæ, Pselaphidæ, Scydmænidæ, Thorictidæ, Rhysopaussidæ, Endomychidæ, Silphidæ, Lathridiidæ, Histeridæ, Scarabæidæ Brenthidæ, etc., which partly represent some myrmecophil or termitophil families or sub-families and partly embrace a larger or smaller number of myrmecophil or termitophil genera and species, whose systematic characters attribute themselves as adaptive characters to the myrmecophil or termitophil mode of existence.

Furthermore we find also in the insect order of the Diptera or two-winged flies a series of similar examples. The Termitoxeniidæ form a special family (or subfamily) which embraces exclusively Termite guests, of which two genera (Termitoxenia and Termitomyia), not only in their entire form of body, but also in their development and mode of reproduction, strikingly differ not only from all other Diptera but even from all other insects. They possess, namely, no larval form, but effect their post-embryonal metamorphosis in the form of a peculiar ‘ imaginal evolution ’ (Fig. 37). Also the individuals are not sexually separated, but are regularly protandric hermaphrodites—a unique case in the insect world. All the morphological and morphogenetic (embryological) characters of this remarkable Diptera—of which so far six species are known—show themselves to be termitophil adaptive characters.

Furthermore among the termitophil Diptera must be mentioned the Termitomastini, which, although belonging to another sub-order of the Diptera—the Termitoxenini are short-horned, the Termitomastini long-horned—present many similarities to them. Also the Termitomastini owe their systematic separation entirely to their adaptive characters fitting them for a termitophil existence.

A very interesting example of termitophil transformation is furthermore shown in the genus Thaumatoxena belonging to the Diptera family of Phoridæ. Therein we find that even the organization characters of the two-winged order are so greatly masked by the termitophil adaptation, that two excellent insect experts, Breddin and Börner, described the first species of this genus *Thaumatoxena Wasmanni* originally—not as a Diptera but as a new genus of the Rhynchotæ—which, however, form quite another order of insects.

We must also mention here the myrmecophil wingless Diptera genera Aenigmatias and Oniscomyia, which resemble rather a Blattid of the order of Orthoptera or a small Isopod than a fly—and this again through the adaptation of characters to the myrmecophil mode of life to which they owe their systematic peculiarities. Also among the other Diptera we meet with many myrmecophil genera, such as Microdon, Ephippomyia, Harpagomyia, etc.

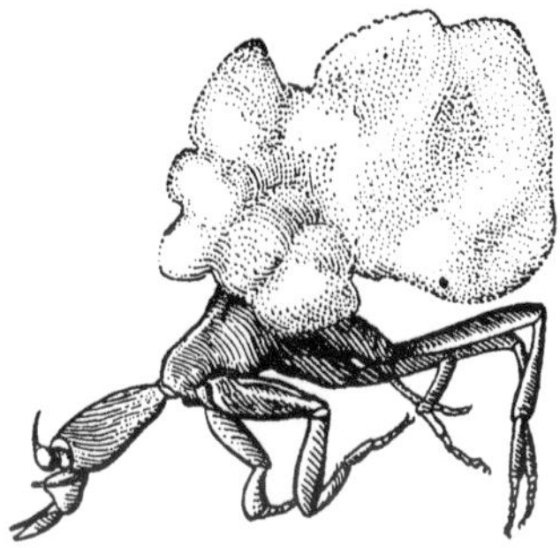

Fig. 37.—Physogastre Imago of *Termitoxenia Assmuthi*. (*After Wasmann.*)

We come now to the underlying principle of the above examples. It is based on the evidence of palæontology, comparative morphology and biology, and the individual evolutionary history.

(*a*) Palæontology shows us that the systematic orders of the Arthropods, to which the Ant guests and Termite guests belong, appeared very much earlier in the world's history than the Ants and Termites themselves. Thus, so we conclude, the guests belonging to those older Arthropod orders could not be absolutely 'created' for their later-coming hosts, but have only later been evolved by way of natural evolution out of originally independent living forms by adaptation to the myrmecophil or termitophil mode of life into the systematic species, genera, and families such as we find to-day.

An example out of the class of insects we give in more detail. The order of the Beetles appeared geologically already in the beginning of the Mesozoic group of formations in the Trias, where it is represented by about twenty genera. Altogether 352 Mesozoic species of beetles have long since been found.[1] The order of the Termites (Isopteræ) appears, however—so far as hitherto can be certainly known—first in the beginning of the Tertiary period, therefore at the commencement of the Cænozoic group of formations : in the Eocene we find first one species, in the Oligocene twenty-five, in the Miocene twenty-nine. The family of the Ants of the order of Hymenoptera, so far as can be certainly known, first appeared in the lower Oligocene—thus in the older Tertiary period: in the Oligocene there are 121, in the Miocene 174 species—which for fossil insects is a very large number.[2] In any case we see from this that both the Ants and the Termites only, in the beginning of the Tertiary period, became a power in the household of the world, which was the essential preliminary condition for the adaptation of other insects to the mode of existence in the nests of the Ants and Termites. We must therefore necessarily assume that the myrmecophil and termitophil species, genera, and families of the Beetles either were first subsequently created in the Tertiary period—and that means each species, genus,

[1] Handlirsch: *Die fossilen Inseckten*, pp. 398, 1171.

[2] *Ibid.* 1182, 1185.

and family of guests for a particular 'normal' host species, host genus, and host family, and this at the same time; or we must assume that they, by the way of a natural racial evolution, by adaptation to the myrmecophil and termitophil mode of existence, arrived at that which they represent in systematic classification—namely, particular species, genera, and families. The first assumption, however, gives no natural explanation whatever of the origin of adaptive characters, but is an obvious denial of the same; therefore the latter assumption remains as the only natural explanation of the facts concerned.

For the further elucidation of this evidence it may be furthermore mentioned that, for instance, the various species of the wandering Ant genus Eciton in Brazil have not only various species, but even very many varied genera of Hemiptera as guests, which are only adapted to this kind of host. This applies namely to the guests of the mimicry type which are extraordinarily specialized. Thus, for instance, the genus Mimeciton is only fitted for existence for and with *Eciton prædator*, the genus Ecitoxenia only for and with *Eciton quadriglume*, the genus Ecitophya (Fig. 38) only for and with *Eciton Burchelli* (Foreli), etc.; and, in addition, the species concerned of the host genus Eciton are very closely related and partly so closely resemble each other that a constancy theorist could only regard them as 'races' of one and the same species and would therefore not require for them a special 'creative act,' but the guests God must have 'created specially.' The futility of this assumption is obvious.

(*b*) The second source of evidence for the underlying principle is to be derived from the facts of comparative morphology and biology, and specially from the examples of recent formation of species which we have derived from a series of works on the Ant guests and Termite guests. We find, namely, still in the present time clear traces of a formation of 'new species' in this

Fig. 38.—*Ecitophya simulans Wasm.*
(S. Catharina, Brazil; mag. 5 diam.)

domain of research. Two of such examples must here suffice to be mentioned.

Within the hemipterous genus of Dinarda we find a series of bicoloured (red and black) 'forms' which live with various species of the Ant genus Formica in Europe and Asia and are so adapted to them that they regulate themselves as regards their size and coloration according to those of their particular host Ant species. Two of these forms—*Dinarda dentata* and *Märkeli*—which live with *Formica sanguinea* and *Formica rufa* respectively, cannot already be discriminated from so-called 'good species,' and occur also throughout the whole domain occupied by their hosts. Two others, on the other hand—*Dinarda Pygmæa* and *Hagensi*—which live with *Formica rufibarbis* and *F. exsecta* are only on the way to develop: they appear, for instance, only in a limited portion of the domain occupied by their hosts and in their typical form, outside of which they are absent or are replaced by transitional forms which have arisen from *Dinarda dentata*; they stand thus as outposts at different stages of species formation, always according to the various points of their extensive domain. In this way we conclude that there is being perfected before our eyes a so-called process of species formation within the genus of Dinarda.[1]

The same causes of adaptation which at present still determine the process of differentiation between our northern bicoloured Dinarda forms, suffice, however, perfectly to explain the systematic differences which exist between the genus Dinarda and the closely related therewith Mediterranean genus Chitosa. The host ant of the latter is namely *Aphænogaster testaceo-pilosa*, thus belonging to quite another sub-family of the Ant stock (Myrmicini) belonging to the Formica (Camponotoni). The adaptation of an Aleocharina of the offensive type—such as are Dinarda and Chitosa to Formica on the one hand and to Aphænogaster on the other—demands, however, much greater morphological differences than the adaptation to different

[1] The objections raised by H. Muckermann in *Natur und Offenbarung*, 1909, No. 1, have already been contested by me therein in No. 6.

species of one and the same genus of host Ants. It is therefore easily comprehensible that the differences between Dinarda and Chitosa can be raised to the value of generic distinguishing characters, while those between the various bicoloured Dinarda forms attain at the highest the value of specific characters. Thereby is the demonstrative power of the argument deduced from the evolution of our Dinarda forms also extended to the genera of Dinarda and Chitosa.

A second example, but certainly of only recent species formation, is presented by the transformation of East Indian and African Wandering Ant guests into Termite guests. Within the hemipterous genera of Doryloxenus and Pygostenus, whose entire generic types are only to be explained by adaptation to the mode of life of Wandering Ants (Dorylus and sub-genus Anomma), there are found namely amongst numerous dorylophil species also a few termitophils which, together with the generic characters which indicate Dorylus guests of the offensive type, show special specific characters, which render them proper termitophil species. Since, however, we can only explain the systematic generic characters of these hemiptera, which belong to the sub-family of the Pygostenini, by adaptation to the dorylophil mode of existence (see above, p. 191), we must logically explain the specific characters which deviate from the dorylophil relatives, and are shown by the few termitophil species as follows, viz.: that these, geologically speaking, in quite recent times, have become transferred from the mode of life with Wandering

Ants to that of the Termites, and have thereby become new systematic species, since they, as a consequence of their new mode of life, assumed characters by which they approach the rest of the termitophil Aleocharinæ of the offensive type of the genera Discoxenus, Termitodiscus, Termitusa, Termitopsenius, etc.

Since, in addition, Wandering Ants by preference attack and plunder Termite nests and on such occasions are also accompanied by their guests of the genera Doryloxenus and Pygostenus—the first riding on the Ants, the latter going on foot—from the biological standpoint it is also easily explained how this transformation of originally dorylophil species to a termitophil mode of life has been brought about.

(*c*) A third proof of the underlying principle of our evidence is seen in the individual evolutional history especially of Termitoxenia. This hemipterous genus possesses, namely, peculiar staff-shaped dorsal growths, which it is true stand on the site of former wings, but are entirely useless for flight, while they serve various other biological purposes, such as feeling organs, transport organs by aid of the hosts, and as exudatory organs. That, however, these enigmatical structures were originally wings, that thus the wingless Termitoxenia have arisen from normally winged Diptera, can still be shown to-day by the individual development of these thoracic attachments; since they still show, in a certain youthful stage of the animal, the form of wing flaps with a clear wing venation, which are later absorbed and changed into the staff-like attachments.[1] We have thus also in the individual development strong evidence of the probability of the evolution of this genus from the original stock.

By the above our concluding principle is also established, viz. that the adaptive phenomena in the Ant guests and Termite guests provide an abundance of evidence of the generic historical evolution of new species, genera, and even families in the animal kingdom—i.e. their evolution from the original stocks.

§ 3. *Suggestive points in the Embryogeny of the present organisms.*

(1) *Premises and extent of the embryological evidence.*

We have seen that even to-day causes are active which can lead to the most multifarious transformations of animals and plants. Thereby is clearly shown that the germ cells of the altered forms are also influenced. We will elucidate this by a simple example. If a plant be transplanted from a valley, its usual habitat, to an elevation of considerable height, there arise various adaptive characters: the plant produces, for instance, hairs as a protection from cold, its leaves become very rich in chlorophyll (intensely green) in order to assimilate more vigorously during the shorter period of vegetative activity; thereby the whole habit of the plant can be

[1] See Wasmann: *Die Thorakalanhänge der Termitoxeniidæ* (*Verh. der Deutschen Zoologischen Gesellschaft*, 1903).

greatly altered. If, now, after some vegetative periods the seed of such purposely adapted 'alpine' plants be sown again in the valley, there is shown in the resulting individuals a strong tendency to retain the 'adaptive characters.'

Obviously in this case the formation of the seed (the embryo) alone was under the influence of the elevated position: the entire growth of the plant itself occurred in the valley. What, therefore, still appears of the alpine character—and that in the first generation is fairly considerable—was established in the seed (embryo) and therefore already in the ovum.

In other words, the entire embryonic development is so fashioned that it no longer strives towards the earlier forms but rather towards the newly acquired adult ones.

This applies exactly to the other cases: if the alterations which mostly appear in the full-grown complete forms have arisen by parasitism, by particular modes of life, by isolation, etc., the embryonic development is always influenced; if it were not, the changed adult forms could not present themselves at all.

The deeper the transformation of the entire organism may be which it needs in order to become adapted to its new mode of life, the greater will its embryonic evolution differ from the earlier one. The more trifling it is, the less also will the embryogeny be altered. Furthermore, there are visible the alterations, in those germ stages in which the organs commence to form, which are specially designed for service under the new life conditions ;[1] for the rest the germ development will proceed as hitherto. If now this unaltered remainder is so constituted that there can be clearly perceived in it a mode of evolution peculiar to a determined type, then we must regard the animal concerned as a member, as a variation of that type. By what causes the deviations in the completely formed condition and in the embryonic stages leading thereto are induced can then under some circumstances be directly seen. If the larva, for instance, attaches itself firmly and at once commences the transformation which strikes us in the perfect animal, then was this transformation formerly caused for the first time by transfer of the animal to a sessile mode of life.

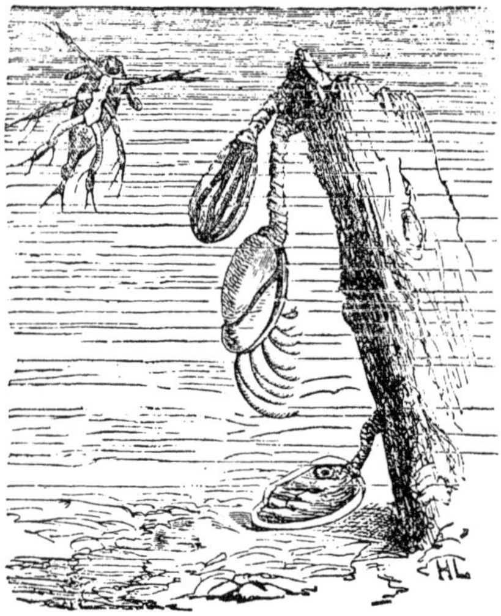

FIG. 39.—BARNACLE IN ITS FORMS OF DEVELOPMENT.

[1] With regard to the time in which the separate tissues and organs became differentiated during the embryonic development, there are often great differences in the separate groups of animals. Biologists define two great groups, the mosaic and the regulation ova, according to the earlier or later commencement of the differentiation and the therewith connected greater or lesser facility of the 'regulation' (restitution) in the embryos produced from the eggs.

(*a*) A few examples will make this more comprehensible. On driftwood and sunken piles there is often found a mussel-like animal attached by a stalk—the Barnacle (*Lepas anatifera*, L.) (Fig. 39). Formerly they were regarded as mussels, although even on a superficial examination much is perceived which is not exactly mussel-like, as for instance the possession of numerous movable limbs, the clasping feet (*Rankenfüsze*).

The most remarkable thing is that from the eggs of this animal crab larvæ issue of the so-called 'Nauplius' type. The Nauplius larva is an embryonic stage common to all the lower crabs (Entomostraca). After a definite time the larva attaches itself by its own tentacles to a support, lime is deposited in the shell, the head becomes a stalk, the eye is aborted, the swimming feet become clinging feet (which serve to whirl food within reach but not for locomotion), and the Barnacle is complete.[1] On close examination we see certainly also, in the adult form, still other true crab-like characters, as for instance in the construction of the mouth, the nervous system, the legs, etc. In short there remains everything of a crab type: in the first place the embryogeny up to that stage where such constructions were added, the positive adaptations to the sessile mode of life, and furthermore all the characters of the Cirripedia, which also can be of service in their old form in the new mode of life—jaws, nerve system, and make of the legs. In other cases certainly it is

[1] Heffe: *Abstammungslehre und Darwinismus*, p. 30.

only the embryonic development which now gives a clue to the systematic relationship—for instance with *Sacculina carcini*, the female of which finally becomes an egg-bag which pervades the whole body of its host animal (a crab) with a cotton-like web. The embryonic development on the other hand is that of the Cirripedia (Figs. 40 and 41). From all this we conclude, and, as

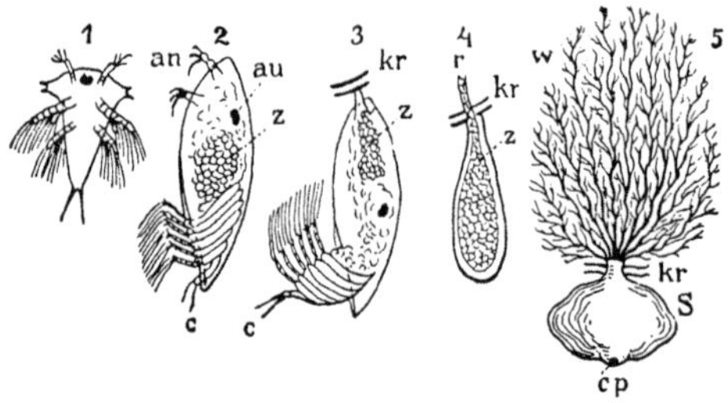

FIG. 40.—CIRRIPEDIA. *Sacculini carcini* (*after Delage*). 1. Second Nauplius stage. 2, 3. The same after attachment of the breast piece and loss of tail. 4. The bottle-like stage passing into the interior of the crab. 5. The final stage.

Letter reference: an, antennæ; au, eye; c, tail; cp, anus; kr, main shield of the crab; r, the cell tube of the *Sacculina* larva penetrating the host; s, *Sacculina externa*; z, central cell mass. (*After von Graff.*)

it appears to us, with entire right, that the Lepas species were formerly free-swimming crabs which subsequently adopted a sessile existence. Their embryogeny alone and the still remaining crab characters show us clearly the true nature of the Barnacles. The whole group of the Cirripedia behaves in a similar manner. Very remarkable beings are also the animals known as Parasitica which as parasites live especially on fish, upon whose skin or gills they attach themselves by suckers. They were formerly considered as worms or articulate animals, until their embryogeny became known. 'They possess an amorphous body in which often nothing of limb formation remains and only a trace of extremities can be found.'[1] Some characters,

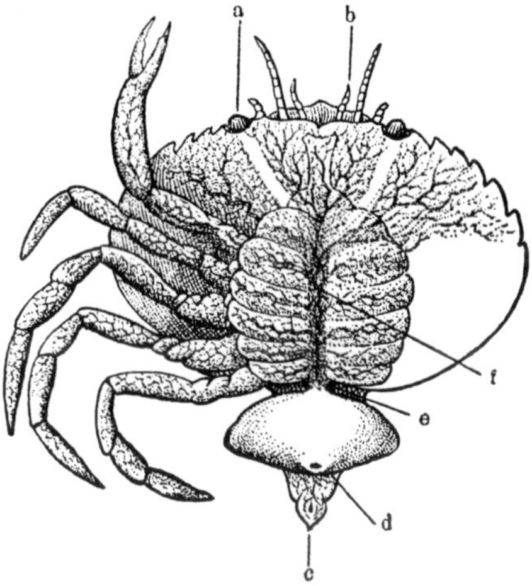

FIG: 41.—*Sacculini carcini*, fixed on *Carcinus mænas*, whose abdomen is exposed. a, eye; b, antennæ; c, anus of crab; d, shell opening; e, stalk; f, root of web, enveloping the intestines of the host, leaving the germinal region free.
(*After Hertwig*: 'Zoologie.')

which remind one of free-living Swimming Crabs (two suspended egg sacs), furthermore a series of transitional forms between Swimming Crabs and these amorphous beings, and above all their evolutionary history (embryogeny)—these were the factors which cleared up their systematic position. They pass through the typical Cyclops stage of the Copepods, and only when

[1] R. Hertwig: *Lehrbuch*, p. 382.

the females, and only the females, attach themselves after the pairing, do the retrogressive steps begin which lead to the assumption of the almost limbless state. The males remain much more crab-like, they die after the pairing, their function being fulfilled ; the females, on the other hand, must now provide the eggs with nutrition, and therefore survive. Since they can do

FIG. 42.—Young Flounder before the shifting of the eye. (mag.40diam.)

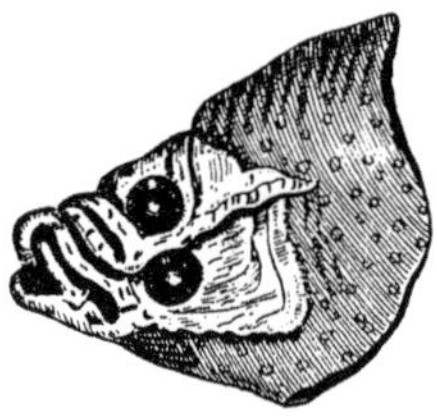

FIG. 43. — Flounder when eye is quite shifted.

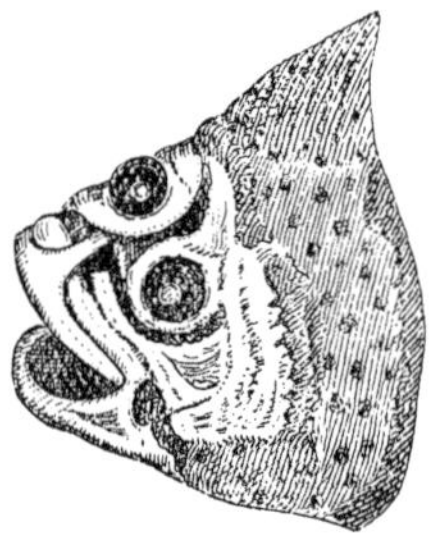

FIG. 44.—Commencement of the shifting in the Turbot. (mag. 10 diam.)

that as parasites without organs of locomotion or sense, these abort as superfluous. They were therefore at first Rudder Crabs (*Ruderkrebse*) in their appearance, as the males still are, and their amorphous form is no original one, but one acquired by parasitism.

Many similar examples might be quoted—as, for instance, the transfer of the right eye to the left side in the young of flatfish which, as adults, lie on the right side and have both eyes on the left side (Figs. 42, 43, 44, 45). If that were originally so, why have the flatfish an embryogeny which strives from the beginning towards this peculiarity? Why are the eyes, as with normal fishes, always singly placed on both sides and travel first of all in the young fish so as to come together?

We have thus learned of some cases in which the individual evolutional history really explains to us how the adult forms formerly appeared.

It is scarcely necessary for us to emphasize the fact that as regards the origin of those types themselves, fish or crab types, we learn nothing at all, but only how some fish and crabs can arrive at a form deviating from the normal. It may, however, be emphasized that everyone who accepts the above conclusion must simultaneously agree that to each type there belongs also a particular process of evolution, otherwise there could not be expected, from the embryogeny, any explanation of the systematic classification.

(*b*) Of somewhat wider application are the consequences of the conclusions which have been drawn from observation of other dissimilar peculiarities of the embryogeny of many recent animals. The Salamander (*Salamandra maculosa*) is viviparous and produces its larvæ in the water. The larvæ possess in conformity therewith gills for breathing water and a rudder tail for swimming. A quite near relative, the black Alpine Salamander (*S. atra*), also viviparous, bears only two to three young, which are born on land; the young are conformably provided with lungs for breathing air and with a round tail for creeping. But these young ones pass through, in the mother's body, a stage with well-formed gills and a rudder tail, which naturally are never of service.

What follows from these examples, and what has been inferred? Other newts, like the Tritons, lay their eggs in water, from which then in the first place there issues a larva developed with swimming tail and gills and later the form adapted for land life of the lizard with lungs and round tail. The two Salamanders above mentioned

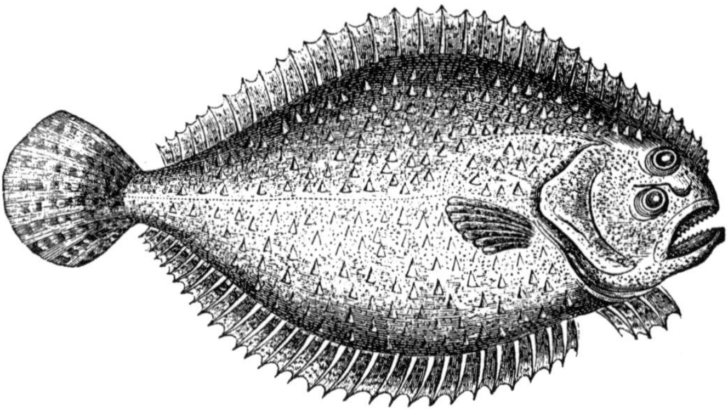

FIG. 45:—TURBOT:

produce first the larvæ, the one kind earlier, the other later.

This therefore implies that to the type of the tailed newt, specially to the sub-order of the Salamandrinæ, a larval form belongs, which lives in the water and is conformably equipped, but that the development, which with all is the same and remains the same, may be more or less intra-uterine. Why that happens we need not really know; for the Alpine Salamander, which lives in damp woods of high elevation, the necessity or also the possibility of bringing forth the young in water in any case ceased to exist. That in the first place did not alter its embryogeny at all, and consequently it still produces the larval form which now appertains to its type. Experiments have also been successful, which is not to be wondered at, in causing it also to produce its young in the water, provided with gills and a rudder tail.

Such vacillations between inner and outer (free) embryonic development are very often met with. The 'smooth shark of Aristoteles' (*Mustelus lœvis*) is viviparous in contradistinction to all other Sharks. There are oviparous and viviparous insects and also lizards.[1]

With regard to the viviparous Hill Lizard (*Lacerta vivipara*) Kammerer states: 'it is normally viviparous; the young, from three to ten in number, are, it is true, often at the moment of birth still enclosed in the egg skin, which, however, in a few minutes or hours they burst open. If, however, the parent animals are kept in an unaccustomed warm temperature of at least 25°C., then they lay eggs, whence the young cannot so quickly escape; at the first egg-laying period in high temperature the time between laying and hatching out is from three to nine days, the eggs are no more numerous than before and have no shell;[2] . . . the

[1] The following details are taken from *Vererbung Künstlichen Zeugungs- und Farbenveranderungen* of Dr. P. Kammerer in the *Umschau*, 1911, No. 7.

[2] In consequence of the heat the Lizards are also of a darker colour, which is recognizable already in the embryo still inside the egg.

second laying period yields us, however, five to twelve eggs, which are enveloped in a parchment-like, yellowish-white, opaque shell, like those which other Lizards possess.'[1]

With the Meadow Lizard (*L. serapa*) Kammerer has established the fact that the normally parchment-like eggs become quite hard-shelled (and at the same time round) if the parents are kept permanently under a temperature of 30° to 35° C. (Fig. 46). If the lizards be restored to the normal cooler conditions, the first generation lays still hard-shelled eggs; and also the young which are born under normal conditions from the 'heat forms,' which have become black by reason of the warmth, still clearly show the black coloration in the first generations and, quite naturally, the more so the nearer they approach the adult form. The very first stages show still clearly the lighter colouring.

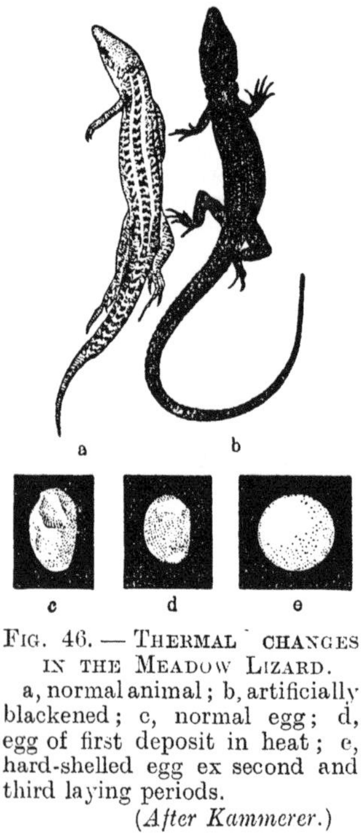

FIG. 46. — THERMAL CHANGES IN THE MEADOW LIZARD.
a, normal animal; b, artificially blackened; c, normal egg; d, egg of first deposit in heat; e, hard-shelled egg ex second and third laying periods.
(*After Kammerer.*)

This example confirms in all points what we have so far said regarding the influence of the external world and the influencing of the embryonic stages; it shows

[1] 'Other' = normal egg-laying forms.

also that there are many vacillations in the relations of the intra- to the extra-uterine period of development.

In the Alpine Salamanders we observe how an internal stage can be omitted; in the Mountain Lizard the contrary can be effected by experiment.

(*c*) We will now assume that the causes which have led to the said change in the Alpine Salamander continued, and also increased in power, so that the inutility of a constructed gill stage became ever greater. We know that in such cases a tendency immediately shows itself—and this is the case with all organisms—no longer to form such non-functional organs. The result will be that at first a defective construction follows, until finally the former organ perhaps entirely disappears, or it may be used in quite another form. The germ development leads ever more directly to the new form, since the previous one, under the altered circumstances, is no longer suited to its purpose.

The gill branchæ of the larval Alpine Salamander could consequently quite well be so far reduced finally that only splits or folds would be formed in the gullet as, with the other Salamander larvæ and gill-breathing animals, precede the formation of the gills as preparatory stages. Then we should have a rudiment in the sense of a formation which, by its construction and position in the organism, has a similarity to definite organs of other animals, but so imperfectly developed, or even only suggested, that they can no longer exercise a function, or at any rate only extremely imperfectly. That would be specially a rudiment of an evolutionary stage.

The entire significance of the rudimentary gill in this case would obviously consist therein that the Alpine Salamander had become incapable of depositing its young in water, as it formerly did. It remained, however, despite this change, in every respect a true Salamander in habit and also in the embryogeny, in which only that was altered which could no longer serve.

Regarding the origin of the Salamander, as member of the Amphibia or Salamander type, the rudiment says absolutely nothing; its explanation as rudiment presupposes rather the existence of the Alpine Salamander.

(*d*) Rudiments now play an important rôle in the theory of evolution, but mostly a very inglorious one. Conscientious research must, however, in the first place ascertain whether a rudiment really exists before conclusions are come to. Wherever an actual function can be determined, or some formation under consideration cannot generally be regarded as an organ, but, for instance, only as a necessary preliminary for the fashioning of the adult form, then there is no question of a rudiment. In the first case the actual function fulfilled explains entirely the existence of that organ, and no ground longer exists for seeking for another earlier function, which the presence of that formation should render comprehensible.[1]

In the second case it is precisely so. The organs

[1] For a long period it was the fashion to designate all formations, whose function was unknown, straightway as rudiments. In this way Wiedersheim has made a large collection of rudiments in man. See, for instance, E. Wasmann: *Biologie*, etc., p. 454, and *Kampf um das Entwicklungsproblem*, p. 94.

arise in the embryo not all at once; they are gradually constructed. A stage, therefore, which can be recognized as such a beginning from its actual subsequent fate, has absolutely nothing to do with a rudiment.[1]

Examples of actual retrogression are not rare. In some whales (e.g. *Balæna mysticetus*—Greenland Whale) there are found entirely buried in the flesh some remains of the pelvis and the upper and under thigh-bone, both imperfectly formed. Externally, of

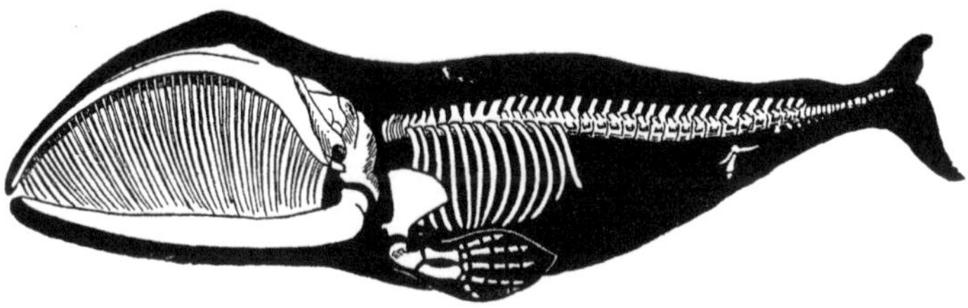

FIG. 47.—BALÆNA MYSTICETUS (Greenland Whale). (*After Heffe.*)

hinder extremities there is nothing now perceptible (Fig. 47).

It is clear that these bones form no longer a functionally capable leg. We assume, therefore, in order in any case to have a reason for their existence, that they are rudiments (remains) of formerly normally constructed extremities which have become superfluous owing to the adaptation of the animal to life in the water and which appear destined to disappear entirely. In other species of whales there is no longer a trace.

The Seals (Sea Lions) show also a rudimentary

[1] A renowned 'rudiment' of this kind we hear of in the gill slits of mammals and man.

condition of the hinder limbs, but in some forms they can still be used (by pushing) for locomotion.[1] Rudiments are also those temporary teeth which appear to fulfil no function, which we note in young whales, in bird embryos or young birds ; also the temporary and defectively formed wing stage of the Termitoxenia, a termitophil fly (see p. 201).

Conclusion from (1).

The actual points deduced from the facts of embryogeny all lead to the conclusion : That we obtain manifold information regarding the former conditions of organisms but none at all regarding the origin of the types to which they belong.

If the nauplius stage of the Lepas species really shows anything, that arises because there is ascribed to the type of the lower Crabs also a typical embryonic development. The Lepas were therefore formerly free-living Cirripedia, which is seen by the still unaltered remainder of the ontogeny and the still recognizable crab characters of the adult form.

The Sole was really a fish when it assumed the prone mode of life, the Termitoxenia a fly when it became a termite guest, the Whale a pure mammal when it 'went into the water,' etc.

How Flies, Soles, Crabs, Mammals have phylogenetically arisen none of the examples show us.

[1] The front extremities are changed into rudders (fins) in Whales and Seals, but the internal skeleton is still fairly that of the quadrupedal Mammalia. They show thus still more of a positive adaptation.

(2) *The speculative utilization of embryological evidence.*

Speculation has not been satisfied therewith. Each 'rudiment' which is met with in the embryogeny of a present organic form should become a document of the actual historical evolution of the type itself, at least in such cases where the rudiment concerned possesses a similarity to a really functional organ of an adult type, even though a very distant one. A 'germ rudiment' of a mammal embryo should no longer point to a former free-living larval stage of the mammal but to a true adult fish or a true tadpole. That is maintained and believed.[1]

The most definite and most general formularization of the embryological argument is given by Häckel in his 'biogenetic fundamental law.'

In the best-known form it runs thus:[2] 'The ontogenesis, or the development of the individual, is a short recapitulation, controlled by the laws of inheritance and adaptation, of the phylogenesis of the ancestors which form the pedigree of the individual concerned.'

[1] It will perhaps be said that 'true' is arbitrarily used; 'fish-like' was only intended. To that it may be replied that the argument, in the form in which we have presented it, is at least used by the Häckelites. It should be really applied to lower classes—for instance, Reptiles or Fish. If the hypothetical fish-like ancestors of the Mammals were no fishes and yet 'true,' then this application does not hold good. Furthermore Häckel himself speaks quite simply of 'ancestors' which may be ascertained by his 'basal principle.' The acceptance, however, of 'fish-like' ancestors is not borne out by actual observation of facts, but in the meantime is only based on a fish-like stage in the Mammalia already existent.

[2] *Natürliche Schöpfungsgeschichte*, p. 276.

In the ideal case it would therefore suffice to carefully observe the succession of all germinal stages in order perfectly to know the desired pedigree. That certainly scarcely ever happens, since, according to Häckel himself, 'mostly, in the ontogenetic succession, much is missing and has become lost, that formerly existed and really lived in the phylogenetic chain of evolution.' 'We are, therefore, in most cases, not in a position to determine all the varied form conditions which the ancestors of each organism have passed through separately by direct ontogeny, but are hindered as a rule by many kinds of gaps.'

The influence of those causes which led to the extinction or the 'falsification' (*Falschung*) of some stages Häckel calls 'Kænogenesis.' The mode of expression is not badly chosen; the principle is easily read; the addition, 'controlled by the laws of inheritance and adaptation,' calms the reader, because it permits it to be supposed that Häckel will know the foundation for his law. The 'Kænogenesis' shows clearly that the application is not always easy: we have to deal with complicated cases.

Despite this there is lacking in the Häckel fundamental principle absolutely all that must be demanded for a scientific principle for the elucidation of the actual genetic history of an organic group. In the first place it is purely an *a priori* assumption and not one based on observed facts. It leads, in the second place, logically followed out, to actually impossible consequences and misunderstands entirely the essence of the embryonic evolution; and finally, it has come not only to no conclusion with the palæontologists, but has led them towards assumptions which the facts directly contradict.

(*a*) The 'biogenetic fundamental principle' is a mere assumption which asserts that which is to be proved, and does so in contradiction to everything that the actual observations permit of assuming.

Häckel says himself that in the embryogeny 'mostly' much is lacking. We can, however, only know whether something is lacking and what it is if, on the other hand, it is clearly established what should be there. That which should be present Häckel must thus have seen in some other quarter, and not in the embryogeny itself; otherwise we should have had a quite indisputable circle of conclusions.

The origin of the organisms—and this by a quite definite chain of ancestry—is therefore assumed here. But whence? We have given above some of the best known examples which told us something regarding the life of the predecessors and their appearance; they all, however, without exception, show only what an organism had to contend with when it was already a member of a well-defined type—for instance a fish, a crab, a mammal, an amphibian, etc. Of the origin of the Crabs, Fishes, Mammals, etc., we learnt nothing at all. Thereby for instance there might be deduced from a rudimentary gill stage a mammal or a formerly free-swimming larva of a mammal, but not that there were formerly no mammals, but only fish.

Thus, from the actual observations of the alteration

in the embryogeny of many organisms, Häckel does not know that the Mammals were once fish or tadpoles; in them, therefore, he has no 'norm' for what is 'lacking' or 'falsified.'[1]

(*b*) The logical extension of the principle leads to quite untenable consequences, or it proves, if these consequences be avoided, nothing more regarding the origin of the types (Birds and Mammals for instance).

A simple consideration will convince us of this. In the individual genesis of a present-day bird there arises, as experience shows, a hard-shell egg stage which is usually termed a 'bird's egg.' This bird's egg is not to be confounded with the actual female germ cell which scientifically is generally termed egg (or egg cell). The germ cell is originally a microscopically small cell which quite early accumulates in the protoplasm yolk particles for the nourishment of the future embryo and thus forms the 'yolk.' This yolk is the fertilizable female sexual cell; only after fertilization, when traversing the oviduct, are further coatings added —the material for which is supplied by glands—viz. the albumen (white of egg), then a fine double egg skin (directly under the lime shell and detachable), and finally the hard lime shell. Thus is the egg finally 'laid.' The germ cell alone, or already in the form of a germ stage, is thus enclosed in the egg when laid.[2]

[1] That palæontology also gives no 'norm' we shall soon see.

[2] An 'embryo' (germlet) is then, strictly speaking, already present if the single cell stage is no longer there—i.e. after the first 'division' (= division of the developing egg). Usually certainly the term 'embryo' is only applied to more advanced stages.

According to Häckel we can and must come to this conclusion: Immediately upon the monocellular stage of the bird parents there followed organisms which, as hard-shelled eggs with yolk, albumen, etc., swam in the sea or lay upon the sand. These ancestors were absolutely nothing else than such hard-shelled formations. How they acquired the yolk, albumen, and the lime shell, and how they generally worked their way out and, later on, swam as fish in the sea—since later they were fish, as the gill rudiment of the bird embryo should prove—that is a very difficult question to answer.[1]

It is certain that no Häckelian draws this conclusion. He would rather refer to the saving clause that the 'laws of inheritance [?] and adaptation' required many things—in short, that the kænogenesis must not be left out of account.

Very probably it would be said that the hard-shelled egg stage which, precisely according to Häckel's disciples, was certainly not always there but has arisen, may be an adaptation in the embryogeny itself. To the further question—In *which* embryogeny?—every one would involuntarily reply: In that of the bird. Anything else cannot wisely be put forward, and the sense of the reply is simply this: an actual bird has, in its embryogeny, adopted this adaptation, since it no longer,

[1] A particularly logical Häckelian could certainly say with perfect right that it is true that as a rule 'much' was altered and 'falsified,' but not the lime-shell stage, according to his conviction, since a 'norm,' which could decide the question, is not given. The 'laws of inheritance and adaptation' also apply and would do so precisely if, generally speaking, no development of the types had taken place.

as formerly, laid its eggs in the water, but transferred the development of the germ to its interior. Since the albumen, the egg skin, the lime shell are all deposited by glands and these are situated in the oviduct, in the oviduct alone could these glands be arranged as an absolutely necessary condition for the formation of the covers and therewith the newly adapted form of development of the germlets (*Keimlinge*). The formation of the glands is furthermore in that case only a purposely adapted means for the new embryogeny when the egg had been previously fertilized in the uterus or oviduct. There were consequently, at the time when that adaptation occurred, males and females, and the seminal cells were introduced by pairing into the genital apparatus of the female. In brief, the birds were at the time of that hypothetical adaptation the same as now. They had, however, teeth in the beaks and also perhaps (all ?) a longer tail: there are some points of support for this in palæontology and embryogeny.

Among the Mammals we arrive at similar stages which, according to their entire nature, must be accepted only as adaptations of the embryonic life itself—and that in the uterus of a mammal, unless it be assumed that the ancestors of the Mammals formerly lived as grown-up individuals in the interior of the maternal body. What, then, were these mothers ?

In short—for the case that in a general way the embryogeny of the Mammals and Birds was not always the same as to-day—the development of the germ of the present Mammals or Birds shows everywhere adaptations which have been brought about by the Mammals and Birds. With regard to the evolution of the type 'Mammal' or 'Bird' we learn nothing.

(*c*) The embryogeny of the Mammalia is precisely as different from the embryogeny of the Fishes as a completely formed mammal is from a complete fish. Just, therefore, as a complete mammal form can under no circumstances be connected with an adult and completed fish, just so can no single stage of germ development (not even in the so-called germ stage) of the Mammals be connected with an embryonic stage of the Fishes.

That is a result at which O. Hertwig has arrived after many years of zealous research, and which Naegeli has already expressed in the renowned sentence: 'In the egg of the hen the species is just as perfectly maintained as in the hen, and the hen's egg is just as widely different from the frog's egg as the hen from the frog. If this appears otherwise to us this is only because in the hen and the frog many distinguishing characters are obvious, while the distinguishing qualities in the eggs lie hidden therein. If the hen's egg did not contain the entire essence of the species, a fowl could not always arise from it with the same certainty.'[1]

The first part of this citation is also found in Korschelt and Herder's[2] well-known textbook expressed

[1] C. von Naegeli: *Mechanisch-physiologische Theorie der Abstammungslehre*, p. 22.

[2] *Lehrbuch der vergleichenden Entwicklungsgeschichte der wirbellosen Tiere*, part I, p. 136.

approvingly and with a detailed confirmation by O. Hertwig.[1] 'From the fact that the ontogenesis of the plant and animal species usually begins with a simple cell stage, the fertilized egg, it has been concluded that all organisms have descended from common unicellular indifferent ancestors: the hypothesis of a monophyletic pedigree has been put forward. How improbable, however, must this appear to us if we start from the point of view above set forth—that according to the ontogenetic causal law the fertilized egg cells of the different species of animal vary in their being quite as much from each other and are quite as good bearers of the various specific differences as are, at the end of their ontogenesis, the perfected individuals upon whose characters we base our animal system.'

The same considerations suggest themselves when it is sought to be concluded, by reason of the 'similarity' of many embryonic stages of the Mammals to those of adult fishes or the larvæ of Amphibia, 'that the Mammals descend from Amphibia or Fishes.'[2] 'The recapitulation theory in the old sense' cannot therefore 'be longer maintained.'[3]

Häckel's basal principle therefore involves a perfect miscomprehension of the nature of embryogeny. The life of a mammal begins with an egg which a parent animal of the same species has formed as an extract

[1] *Allgemeine Biologie*, p. 674.
[2] *Ibid.* p. 675.
[3] *Ibid.*

of its entire essence. The parent animal produces, however, no amœba but reproduces its own form. Each germ stage is through and through a mammal in the making, devoted to that one object by the preceding stage and itself determining the succeeding one to the same end.[1] The types have thus experienced no greater alteration in their embryogeny than in their grown condition, which in all cases is only the result of germ stages determined in one particular direction. Since, however, the types to the eye—whether according to present systematic classification or according to palæontological finds—as adult mammals never show any association with any other class whatever, therefore also the embryonic stages can present no approximation since they belong to the complete condition and produce it.

(*d*) As regards the applicability of the 'biogenetic basal principle' in palæontology Zittel stated in 1895 at the International Geological Congress: 'If palæontology be consulted, it must be recognized that this hypothesis has not been confirmed in any way.' He then shows by several examples at what 'peculiar ancestors' we must arrive according to the 'basal principle'—for instance, for the Crinoids (Hairstars) and Sea Urchins, which, however, 'would not accord in the remotest degree with the facts.' Such examples

[1] All this has been determined by modern biologists by observation and experiment. See, for instance, Driesch: *Philosophie des Organischen*, I, p. 76; *Analytische Theorie der Formbildung*, Korschelt and Herder, part I, p. 81; *Das Determinationsproblem*; O. Hertwig: *Allgemeine Biologie*, p. 572, '*Die Theorie der Biogenesis*.'

showed the 'futility of the conclusions' which are due to the embryological methods.[1] 'The examples might easily be multiplied tenfold.'

According to Depéret 'this law requires to be applied every time with the most extreme care. It would in no case be able to dispense with the subsequent proof which is provided by the actual evolutionary history, i.e. by the knowledge of the palæontological records.'[2]

That, in ordinary language, means that it might *by chance* so happen, as Häckel's once 'fundamental law' demands, but ordinarily it is otherwise, and in that case the investigator adheres to the other cases. If, despite this, the said investigators grant that earlier forms in the adult condition were permanently so constituted, as their present successors are only temporarily constituted in their 'embryonic' or, better, 'young' stages, that is by no means surprising.

'The palæontological Belinuridæ'—says, for instance, Zittel in his exposition—'resemble a thousand young larvæ of the living (recent) "Swordtail" (*Limulus*); the Pentacrinus, larvæ of the Antedon, stands closer to many fossil Sea Lilies than does the adult animal. Certain fossil Sea Urchins retain permanently the young characters . . . of their still living relatives. The splendid investigations of Hyatt, Würtenberger, and Branco, have shown that all Ammonites and Ceratites (Cephalopods to which the Cuttle Fish belong) pass

1 Depéret: *Umbildung der Tierwelt*, p. 107.
2 *Ibid.* p. 40.

through a Goniatite stage and that frequently the internal convolutions of an Ammonite in their form, ornamentation, and lines of suture [1] resemble any other earlier existing genus in the adult condition.' [2] That is quite natural.

It is quite natural, since each alteration, whether retrogressive or progressive, influences the entire individual. If, for instance, a Sea Lily assumed another form—either by specialization or in a definite direction as adaptation to a definite mode of life—it would therefore only reject the hitherto embryonic development in so far as the new complete form to be created rendered necessary. The new constructions would be simply added to the former constructive process, and so it has remained until the present. The newly acquired is always the last in the ontogenesis, the old becomes always the younger or more embryonic. All cases brought forward and really observed thus only show how little since the Palæozoic period the types like the Sea Urchin, Sea Lily (Crinoids), etc., have changed. The entire transformation indeed is confined, for example, in the Belinuridæ mentioned, to those few alterations which in the evolutionary course of the recent Limulus follow upon the Belinuridæ stage, precisely as in the case of a retrogressive transformation—for

[1] The long since extinct Ammonites (Ammonhorns) were shell-dwelling Cephalopods which in certain interspaces sometimes constructed a new dwelling chamber in the earlier ones; the line of separation between each two chambers is called the suture line (division seam). This forms one of the discriminating specific characters and was subject to constant modifications.

[2] Depéret: *Umbildung der Tierwelt*, p. 107.

example, as the result of parasitism—the entire alteration shows itself between the first abnormal germ stage and the completed form.

§ 4. *Summary of the collected results.*

(1) We have seen how manifold are the alterations to-day which underlie animals and plants. Direct influences (stimuli) by climate, constitution of soil, etc., isolation and close interbreeding through long periods, adaptation to quite special modes of life such as Parasites or Symbionts—all these effect changes in a progressive or retrogressive sense, i.e. in the shape of new formations and more marked differentiation or of regression of existing features.

(2) None of the transformations observed or easily to be regarded as such ever go so far that the allocation to a certain type can no longer be made with certainty, whether in the completed adult condition or in the embryogeny, which indeed likewise forms a constituent of the typical distinguishing character. There arise, it is true, new species, genera, and even families, but no animals and plants with an entirely deviating plan of construction and higher total organization. With the Parasites there remains at least a portion of the embryogeny unchanged.

(3) The alterations which we can determine in fossil organisms are of the same kind as the present ones—i.e. they also carry in themselves the distinguishing characters either of a direct adaptation to the environ-

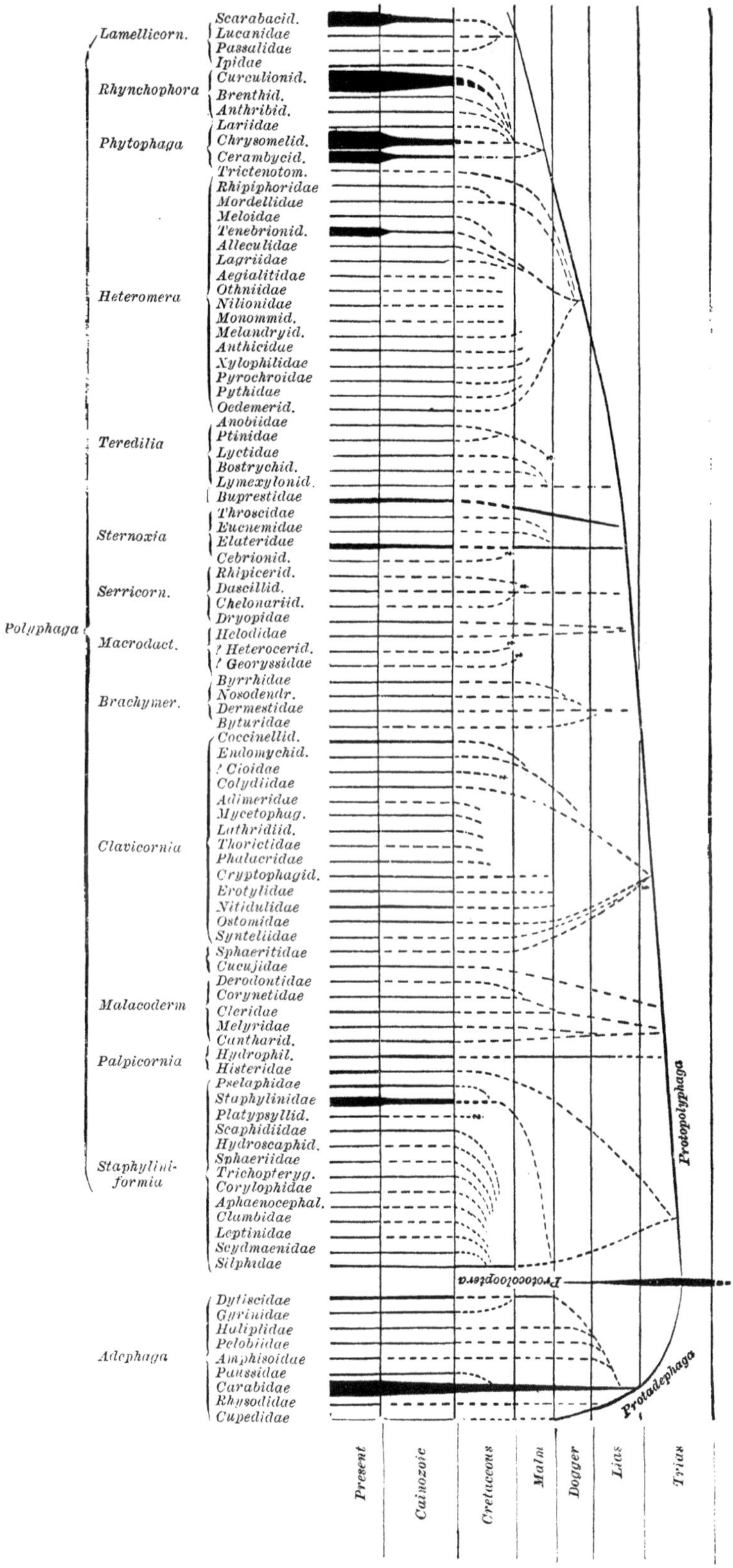

FIG. 48.—PEDIGREE OF THE BEETLE FAMILY (*after Handlirsch*).

ment and quite special objects, habitats, or stations (law of specialization and convergence), or to parasitic and sessile modes of existence (law of regression).

(4) This perfect agreement in the mode and manner of the transformations and their extent, as between the recent and the fossil organisms, shows that the same causes which are busy to-day in alteration of form were so formerly and no others, since otherwise the mode and manner of the transformations could not be the same in both cases.

That, however, the extent of the changes is the same now as formerly—i.e. that they never completely wipe out a given type—shows clearly that a more thorough-going transformation and alteration is excluded. That which has never happened can, according to the principles of natural science, also not happen.

The organic kingdom therefore forms no unit but, as established by natural research, a definite number of true types, i.e. grades of perfection. Ever more and more do the investigators see that their chief task consists therein, to ascertain the history of the separate groups of animals and plants and to discover the laws and causes of their evolution.

How imperfectly even this limited task has so far been fulfilled is shown, better than by a detailed presentation, by the table given as an example (Fig. 48) from Handlirsch [1] showing the pedigree of the Beetle family, i.e. of a subordinate systematically classified category. The extended lines show actual discoveries—they

[1] *Die fossilen Insekten*, p. 1279.

run almost without exception parallel—the lines which should connect the separate families are, almost without exception, broken, i.e. hypothetical.

Conclusion.

The period of 'fantastic evolutional histories,' as Depéret [1] expresses it, or the 'methods of approximating valuation,' is certainly disappearing. For true progress the *a priori* hypotheses of Darwin and others have yielded not only absolutely nothing, but done much harm. Professor Steinmann expresses himself in that connection as follows—in which bitterness is evident since the current evolutional hypotheses have driven him almost to 'despair': 'When a scientific branch of such predominant importance as the theory of descent gets off the proper track it naturally detrimentally influences all the branches of knowledge with which it is organically associated. So it is also with palæontology (and to a certain extent also with geology), which, instead of having an independent basis, has become a vassal of the Darwinistic-Häckelistic theory of evolution. With the low position in which palæontology still remained in the years 1860 and thereabouts, it became at first entirely taken in tow by them; the significance of the formation of species and subjection to the struggle for existence of the phylogenetic meaning of the systematic categories of the unity of origin of the smaller and

[1] *Umbildung der Tierwelt*, p. 143.

larger animal and plant divisions were brought, without proof, into the area of fossil material. No wonder then that palæontology could not follow these academical prescriptions, and, when it tried to do so, made a fiasco.'[1]

The embryological methods of Häckel have, according to Depéret,[2] led the whole of palæontological research in a wrong direction. The 'naive' pedigrees constructed according to them 'have crumbled just as speedily as they have arisen; they cover, as with rotten wood, the ground of the forest and only render more difficult the progress of the future.'[3]

All the more is it to be regretted that the neo-Lamarckians wish to endeavour once more to solve the problem of evolution deductively, since they deduce the common origin of plants, animals, and man from the entirely wrong assumption of their essential equality. Despite all protests there is thereby substituted another 'dogma' in the place of the Darwinian 'dogma' as they express it. The tone of their writings is also not always a high-class one. The investigators who believe in God are contumeliously pitied: thus A. Wagner[4] says of Wigand that his 'in many respects excellent adverse critique of Darwinism, he has spoilt, particularly, through the marked theistic colouring of his philosophy. Like Wigand, too, did K. E. v. Baer spoil the influence of his arguments by deriving from his

1 G. Steinmann: *Die geologischen Grundlagen der Abstammungslehre*, p. 17.

2 *Umbildung der Tierwelt*, p. 113.

3 *Ibid.* p. 108.

4 *Geschichte des Lamarckismus*, p. 60.

view of the world a theistical conclusion.'[1] In a like hostile fashion write also Pauly and Francé.

How, however, without a theistical assumption we can understand the origin of life, the origin of animals and plants, the graduations within the two kingdoms, the faculties for adaptation of the organisms, the tendency to people the air, earth, and water, etc., without intervention of a super-mundane cause, Wagner certainly does not show us.

If we know of the organisms that they can maintain themselves in construction and function in agreement with altered conditions of life, that is nothing more than a statement of the fact but no explanation of it. If we read that animals and plants occupy the air, the water, and the dry land, and arrange themselves accordingly, that also—if it be true—is again only a simple statement. Or have air, water, and land the tendency in themselves to become inhabited?

Let us, however, go into details and ask, for instance, how it comes about that the lark rises singing into the air, many flat fish lie on their sides, some plants become carnivorous, why the plant *Duvaua dependens* produces for the moth, *Cecidosis eremita*, a gall with a circular cover which renews itself on the inner side and is precisely large enough to let the moth escape, etc. If it be assumed that all this was not always so but has

[1] *Umbildung der Tierwelt*, p. 79. When Wagner speaks of scholastic philosophers one would think that he had to do with a host of highly primitively organized thinkers who had not at all attained to a proper 'intellectual organ.' Just as well might Wagner term 'scholastics' all who oppose criticism.

been evolved, then does the question again, as always, recur: Why and wherefore has it been evolved? What need is there for the plant to keep and cherish a moth—since it only does so by constant expenditure of nutrition—and to shape a cover at the right time, not earlier and not later, so that when the moth creeps out of the gall the chrysalis skin and that alone is torn off?[1] We can only say that it must and should happen just so.

Depéret says appropriately:[2] 'In the time in which we live it would be very thoughtless to maintain that we satisfactorily know the general law which has governed the unceasing transformations of organic life from its beginning on the earth to the present day. Neither the mechanical process of a physiological adaptation, nor the immediate influence of the environment, and still less the struggle for existence, permit us to give a suitable, sagacious, and perfect explanation of the magnificent picture presented by the palæontological history of evolution. In this evolutional history there are certainly, without in itself speaking anywhere of the first origin of life, enigmatical points and important facts in existence whose explanation eludes us.' We have, it is true, for many of these 'enigmatical points,' catchwords which have become very popular, but with catchwords alone no problem is ever solved. In the

1 See the excellent article, '*Ein Wunderwerk der Pflanzentechnik*,' by H. Dieckmann, S.J., in *Natur und Kultur*, 1911, p. 485.

2 *Umbildung der Tierwelt*, p 114. If Depéret himself speaks in various places of the 'mechanism' of the evolutionary process, he always means thereby only the external course of development.

first decades of the last century an investigator who desired that his works should be regarded and read as scientific had to write in the style of the Hegelian philosophy, somewhat as Reichenbach did.[1] 'As the area of a circle is not merely centre-point and circumference but also the relations of both to each other, so in nature is everything directed, as in the thinking mind, by thesis (precept, centre-point, unity), antithesis (contradiction, periphery, multiplicity), and synthesis (equalization, combination of the contradictions, circle area, formation).' 'The green plant kingdom arose in the Algæ from the water—we saw these in progressive formation, always maternally shaped only into beautiful types until, soaring into the element of the air, the Moss by anthers limited their forms. The Ferns take up into themselves the budding of the Algæ and Mosses, and, seeking in vain to find a centre by their circination, Cycas and Zamia achieve this finally by forming the acrogenetic axis of growth.'

'Central formation begins from this point as the type of the plant developing itself first from the primary bud of "Isoetes." But the node divides and repeats itself, and intermediate growths extend themselves as stem and scales—as phantom leaves—emerge from the node, and the male is born and for him the primary female. The scales require a stalk and proclaim the flower in the trinity of the realm of plants, etc.'

'Empty words' we may say, but the expressions

[1] From A. Kerner v. Marilaun: *Pflanzenleben*, II, Leipzig-Vienna, 1891, p. 591.

'spontaneous generation,' 'natural selection,' 'sexual selection,' have, as general explanatory principles of the organic world, just as little meaning. They too will disappear, just as the thesis, antithesis, and synthesis have disappeared.

Theories of evolution will remain, since everything points to the fact that there was and is an evolution of the organic world. This evolution, however, does not express itself in quite impossible spontaneous 'leaps' from the inorganic to the organic, or from plants to animals, and also not in plan and objectless hither-and-thither variation, but in a constant maintenance of the harmony between construction and function and the external conditions of life, and in constant development of the bases, since 'bases'—and these, too, for one definite end—must exist, as the result is always in one definite direction—viz. the purposeful, the vitally capable.

Neither was life acquired by the organisms themselves, nor were the evolutional tendencies: both were received from another source—from the Creator.[1]

INDEX

Spottiswoode & Co. Ltd., Printers, Colchester, London and Eton.

For Product Safety Concerns and Information please contact our EU
representative GPSR@taylorandfrancis.com
Taylor & Francis Verlag GmbH, Kaufingerstraße 24, 80331 München, Germany

www.ingramcontent.com/pod-product-compliance
Ingram Content Group UK Ltd.
Pitfield, Milton Keynes, MK11 3LW, UK
UKHW022156190926
475702UK00003B/48

* 9 7 8 0 3 6 7 1 5 0 3 5 8 *